I0474954

Excel

Master

The Complete 3 Books in 1 for Excel - VBA for Complete Beginners, Step-By-Step Guide to Master Macros and Formulas and Functions

Excel VBA

Step-by-Step Illustrated Guide for Complete Beginners to Master VBA

Excel Macros

Step-by-Step Illustrated Guide for Complete Beginners to Master Macros

Excel Formulas and Functions

Step-by-Step Illustrated Guide for Complete Beginners to Master Formulas and Functions

Introduction

Getting Started!

Hello, there future Excel Programmers!
Thanks for viewing this book. This book has been designed to be your go-to book for excel programming. It is drafted in simple English to make your programming experience fun and easy. It's fine even if you don't have the slightest idea about Excel VBA, this book will help you get a grab on the programming in no time. (With just a *little* effort) Before going further, I would like to tell my prospective readers that who exactly is the target audience for this book. Please find below my assumption about you:

✓ You do have an access to a computer (A laptop or a desktop). The computers, in turn, have a connection to the internet.
✓ You are a frequent user of the Microsoft Excel tool

What is Excel VBA?

I am sure you must be excited to jump into the bandwagon of the Excel Programmers. Well, hold your horses a little for now, as this chapter will first give you some critical background information that will help you become an fantastic Excel Programmer. The term VBA stands for *Visual Basic* for *Applications*. This programming language was developed by Microsoft Company. VBA is the tool that is going to help us control and customize the functionalities of the Excel. Throughout this book, you will get to see the term *"Macros"* a lot as well. Before you confuse yourself, let me explain what macros is. The codes that are written to perform operations in excel is known as Macros. VBA, however, is the programming language platform where Macros are written. This is how VBA and Macros are interlinked. The VBA tool can be convenient to perform thousands of different tasks. Here are a few scenarios:

- ✓ It can help in analyzing data
- ✓ It can help us in forecasting and budgeting of data
- ✓ Automating our reports to save up time and improve efficiency
- ✓ Conveying data into visuals such as charts, graphs etc

them, the Word doc should look like the PDF file. I can go on and on with the tasks it can perform, but hopefully, you do now have an idea about its functionalities. In a nutshell, VBA tool helps to speed up the operations performed on excel. Let's take an example, every day you get an excel sheet containing a list of employees with some necessary information. You have to manually perform some formatting like to make the name bold, make the phone number right aligned, and add some colors to the rows. Now if the list of employees is long so doing it manually every day can be an extremely tedious task. This is basically where the VBA tool comes in handy. You can write a macro code regarding all the formatting steps, place a button on your sheet and finally bind the macros with the button. Just one hit and you're good to go.

* * *

Jumping into the VBA Tool

Before we jump deep into the swimming pool, learning to swim should be our topmost agenda. This section is going to give you a feel of the entire VBA tool. It will give you a good grab on the basics.

To get started on the VBA tool, we first need to access the most crucial tab: Developer. Follow the steps to access the tab:

1. Open the Excel tool
2. Right Click on any area of the Ribbon and then select the customize option

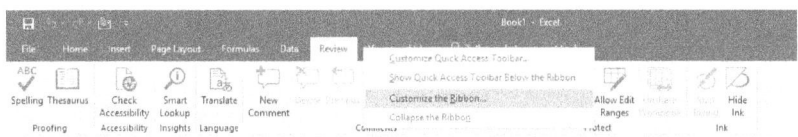

3. Once the customize tab is open, you can find the developer option in the second column
4. Check the Developer Option

5. Click Ok

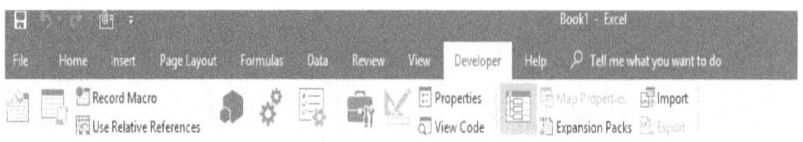

Congratulations! You now have a brand-new tab on your Excel tool: Developer. By clicking on the developer tab, we will be navigated to the section that is of interest to all the VBA programmers.

We will now be performing our very first exercise on the VBA tool. The macro that we are going to write will perform the following functions:

✓ Type name into a cell
✓ Enter the current date and time into a cell
✓ Bold the text entered in the name cell
✓ Change the font size of the date cell to 14

* * *

Recording the Macros

In this section, we will be performing an exercise to learn how to record the Macros. Excel has a built-in functionality in which the user doesn't have to manually write the code. The tool has the option to record all the operations that you are performing and then finally generate a macro code for it.The macro we will write now won't win us the first prize in a VBA programming competition but to reach your final destination one needs to take baby steps. Following are the steps that we will follow to write our very first code.

1. Open the Excel Tool
2. Click on the developer tab
3. Select a cell on the excel worksheet. Click on any cell
4. On the developers' tab, click on the Record Macro Button
5. As shown below, the Record Macros box is displayed
6. Now we will be entering a name for the Macros. The default name is Macro 1 but its always advisable to give a better name

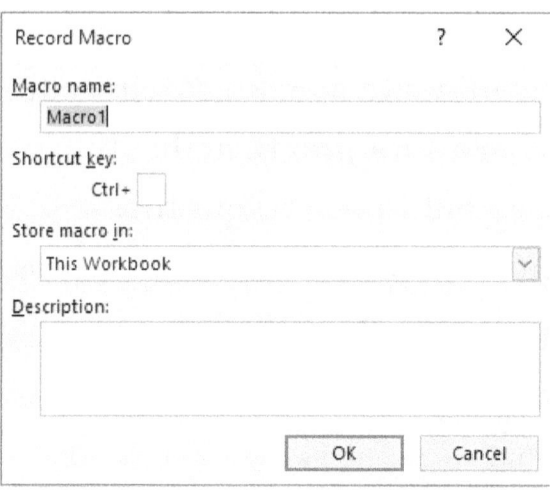

1. Click in the shortcut key and enter a shortcut key. For now, we will be entering shift+N so our shortcut key will be ctrl+shift+N

2. On macro dialogue box, the option that says "store in:" should be set to "This Workbook"

3. Click Ok. The macros will now start to record.

4. Type your name on the cell where you want to display it

5. Now in the adjacent cell, enter the below formula

6. Now we will convert the formula to its value. For this right click on the date cell, select copy. Now right-click on the cell again, and select paste values(V).

7. Now Select the cell where you entered your name. Go to the home tab, traverse to the font section and select bold(B). Also, change the font size to 14.

8. Now to stop recording, go to the developer section. Select stop recording.

   Cheers! Your first project for the Excel VBA macro is finished.

   To ensure that your macros are working fine, you should test your code. Move the cursor to an empty cell, press Ctrl+Shift+N. Within a snap, the excel code will be executed. Your name, date and time will be displayed on the sheet. You can also view the code that has been recorded by the excel tool. Go to the developer tab, and click on macros. You can now view the code that was automatically generated behind the operations that you just performed.

Below is the code that was generated.

```
Sub MyName ()
|'
' MyName Macro
'

'
    Range ("B2").Select
    ActiveCell.FormulaR1C1 = "Anum"
    Range ("C2").Select
    ActiveCell.FormulaR1C1 = "=NOW()"
    Range ("B2").Select
    Selection.Font.Bold = True
    With Selection.Font
        .Name = "Calibri"
        .Size = 14
        .Strikethrough = False
        .Superscript = False
        .Subscript = False
        .OutlineFont = False
        .Shadow = False
        .Underline = xlUnderlineStyleNone
        .ThemeColor = xlThemeColorLight1
        .TintAndShade = 0
        .ThemeFont = xlThemeFontMinor
    End With
End Sub
```

The first statement recognizes as the sub procedure and gives the name of the macro that you entered. The second statement identifies that cell "B2" is selected. B2 means a second column and second row on the excel sheet. The third statement then highlights the name you entered on Cell B2. Cell b3 is then selected and then the Now()

formula was entered. Selection.font.bold= true indicates that the cell formatting was changed to bold. The sub-procedure finally ends the code by the End sub Procedure.

Chapter 1: VBA Variables

Value holders

Let me introduce you to real programming language element called variable. Just like other programming languages, VBA has elements common to them. The main agenda of the VBA tool is to manipulate data. VBA can store the data to the computer's memory. Some of the data that you create is stored in variables. A named storage location in the computer's memory is known as a variable. Excel gives us the flexibility to name the variables whatever we want them to be. To assign a value to the variable, we use the equal sign operator. Let's look at some of the examples of the variables.

```
a = 1
PerformanceRate = 0.95
EmployeeSalary = 123455
DataEntered = False
a = a + 1
UserName = "Anum Haroon"
```

There are specific rules when declaring variables in the VBA tool:

✓ The first letter of the variable should be a character.   The rest can be numbers, letters, or punctuation characters
✓ There is no distinguishment between the upper-case and lower-case letters when declaring variables
✓ A variable name should be without any space
✓ Special characters such #, $, %, &, or! cannot be used in the variable name
✓ The size of the variable name should not exceed 255 characters

To better understand the variables, most programmers use mix cases, for example, PerformanceRate or Employee_Salary.

The VBA tool also puts some restriction on the variable names. Words such as Dim, End, With, Sub, Next and For cannot be used by the programmers as these words are reserved by VBA. If any of these words arise in your code, you will get a compile error.

* * *

Getting Familiarize with the VBA Data Types

When I say the word data type, I am mainly referring to a way in which the program written in VBA stores data in memory. VBA gives us the leverage to not to assign a data type to every variable, but if you leave it on VBA there's definitely some cost to it. The automatic assignment of data types may result in slower execution and wastage of memory. Applications that are small in size may not be affected by the automatic declaration, but the large and complex application does have an impact on them. It's always a good practice to declare the data type of the variables that you are going to use. Although the list is long, below are the most common types of data that VBA can handle.

Data type	Bytes Used	Range of Values
Byte	1	0 to 255
Boolean	2	True or False
Integer	2	-32768 to 32768
Long	4	-2,147,483,648 to 2,147,483,648
Double	8	-1.79E308 to -4.94E-324 for negative values
Currency	8	-922,337,203,685,477 to 922,337,203,685,477

Date	8	1/1/0100 to 12/31/9999
String	1 per char	Varies
Variant	Varies	Varies

A good rule of thumb is to use the variable with the smallest size but it should also serve your purpose.

* * *

Declaring Variables and their scope

By now, I am sure you must be familiarized with about variables and their data types. In this section, we will now be declaring a variable to a data type. if a variable is not declared in your macros, the VBA itself will define it to default data type: Variant. For example, if a particular variable is by default set to be a variant, and it contains a text string that looks like a number (eg 123), this variable, however, can be used for both the numeric and string calculations. As mentioned earlier, it is always a good practice to declare the data type of your variable else the VBA tool will assign the data type as variant and the variant results in time-consuming checks and this ultimately uses memory. One good way to force yourself to declare the data types of the variables is to write the following statement as the first statement in your VBA module:

```
Option Explicit
```

This statement forces you to declare all data types of your variables, as it will throw a compile error if any variable is left undeclared.

A variable is using declared by using the Dim statement.

Let's look at some examples:

```
Dim Employee_name As String
Dim Employee_Salary As Long
Dim X
Dim Amount As Double
'
```

All the variables in the above example have been defined a particular data type apart from the variable "X". The variable X will, however, be treated as a variable.

Apart from Dim, to declare variables we can use three other keywords:

✓ Private
✓ Static
✓ Public

Let's look at an example to get a good picture of the declarations of the data type of variables.

```
Sub Variables()
Dim Name As String
Name = "Anum Haroon"

Dim Age As Integer
Age = 21

Dim Birthdate As Date
Birthdate = 19 / 11 / 1991

MsgBox "Name is " & Name & Chr(10) & "Age is " & Age & " And Birthdate is " & Birthdate

End Sub
```

When the above VBA code is executed, a
message is prompted on the screen. The
message box is as follows:

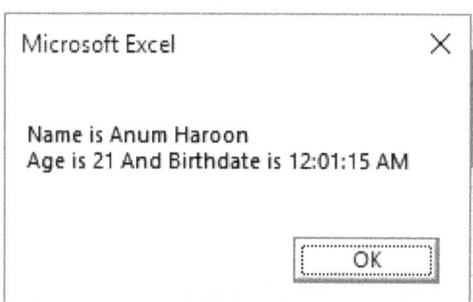

Chapter 2: Strings

Playing with characters

Now let's have fun by playing with the strings in the VBA tool. Strings fundamentally compromise of characters that are arranged in a sequence. The characters in the string can consist of alphabets, special characters, numbers or any of these. The characters in the string must be enclosed with double quotes.

There are numerous functionalities of the string. The table below displays some of the functions that are performed on strings.

Function name	Returns
Str()	*A string representation of a number*
Val()	*A numerical representation of a number*
Trim()	*To remove spaces in a string*
Left()	*To get a portion of the string from the left side*

Right()	*To extract a portion from the right side*
Mid()	*To extract any part of the string*
Len()	*To retrieve the number of characters in the string*
StrCov()	*To convert the string to some other format*
UCase()	*Convert all characters to uppercase*

LCase()	*Convert all characters to lower case*

The Excel VBA handles two types of strings:

✓ Fixed-Length Strings
✓ Variable-Length Strings

As far as the fixed-length strings are concerned, they mainly contain a fixed number of characters. The maximum number of characters they can store is 65,526. The second type of string is the Variable-Length String. The variable-length string can store a massive number of characters. Whenever you declare a string, it's always a good practice to declare the maximum length of the string otherwise the VBA tool will handle it on its own. Below is an example of declaring a fixed length string.

```
Dim MyName As String * 20
```

Let's write a macro code and play with some functionalities of the string. I would like to call the procedure as "Play_With_Strings".

```
Sub Play_With_Strings()

Dim Name As String
Dim FirstName As String
Dim LastName As String
Dim SpaceLoc As Integer

Name = "Anum Haroon"

SpaceLoc = InStr(1, Name, " ")

FirstName = Left(Name, SpaceLoc - 1)

LastName = Mid(Name, SpaceLoc + 1, Len(Name) - SpaceLoc)

MsgBox (" The full name is " & Name & ", the first name

is " & FirstName & " and the last name is " & LastName)

End Sub
```

Now let's decode it step-by-step. The first four statements in the procedure are the part where the variables are declared. All the first three variables have been set as strings and the last one has been declared as an integer. Now after the variable declaration section, we begin by assigning values to the variables. I first stored a name in the Name variable. Now next I wanted to find out the location of the space between my first name and the last name. Space is located through the InStr() function. The InStr() has three input parameters. The first parameters decide from where position should the function start searching in the string. The second parameter is the string that needs to be traversed, and finally, the third parameter defines what needs to be searched. In this example, in the variable value "Anum Haroon", we want to search the location of the space. Once this is found out, the value is stored in SpaceLoc variable. Now we want to retrieve the first and last name and then store in the variables respectively. To do this we will be using the Left() and Mid() function of the string. The Left() function has two inputs. The first input is the string whereas the second input defines the position till where the string needs to be extracted. So by passing the SpaceLoc and the Name variable we can get the first name. To get the last name, we will be using the Mid() function. The Mid() Function has three inputs that are the string, the start, and the end position. We will be passing the name variable, the SpaceLoc position and then the end position which is basically equal to the string length. The last statement is written in the VBA Code to generate an output on the screen. We will again be binding the VBA code to a button. Once the button is executed, the following values will pop up on the screen.

The full name is Anum Haroon, the first name is Anum and the last
name is Haroon

OK

Not too bad prospective Programmers! You now
must be having a good understanding of how you
can play with the string functions. String function
can be convenient especially when you have to
manipulate large datasets. Instead of retrieving
all the functionality manually you can write a
macro code, bind it with a button and you're good
to go.

Chapter 3: VBA Macros

Jumping to an advanced level

So far, we have covered some basics of the VBA Macros. We will now be pacing up our game by jumping over to an advanced level.

VBA Macros has considerably made the lives of excel users easy. With the implementation of VBA Macros in your worksheet, you can significantly save up time by automating your operations of excel. Instead of working on large data sets manually, formatting thousands of records, applying formulas to manipulate data, you can write a VBA Macro, bind it with the excel sheet and do all the operations within seconds.

In the section, we will be working on some examples of VBA Macros. This section will compromise of the following subtopics:

- ✓ Worksheet Ranges
- ✓ Exploring some Excel Settings
- ✓ Pacing up your VBA Code

* * *

Worksheet Ranges

The VBA programming has got a lot to do with the ranges on the worksheet. The following points need to be kept in mind whenever we are working with range objects:

- ✓ If you are not associating the worksheet name to the range in your code, then you must ensure that the sheet on which you want your macros to run must be in the active state

- ✓ Excel gives the users the ability to select multiple ranges

- ✓ The Macro recorder doesn't always create the best code. You can always edit the Macro generated code to make it more efficient

One of the most frequently used operations of the Excel Macros is copying a range. If you are using the macro recorder to generate you a code while copying a range, you will get the below code.

```
Sub Range()
  Range("A1:A5").Select
  Selection.Copy
  Range("B1").Select
  ActiveSheet.Paste
  Application.CutCopyMode = False
End Sub
```

One other operation of the range function is the Selection ability. We often want to select a block of cells and then do some operations on it. Instead of giving reference to each cell, we can collectively refer the entire block of cells in one statement. Please find below an example of the Selection operation on the range statement.

```
Sub Select_Range_Down()
  Range(ActiveCell, ActiveCell.End(xlDown)).Select
End Sub
```

In the above code, the VBA procedure is beginning to select the cells from the active cell. It is then extending the range until a blank cell arises.

```
Sub Select_Range_Down()
  Range(ActiveCell, ActiveCell.End(xlDown)).Select
End Sub
```

One other exciting feature of VBA Macros is the selection of the entire column or the entire row. The following is an example of the selection of the entire row.

```
Sub SelectRow()
  ActiveCell.EntireRow.Select
End Sub
```

Similarly, the entire column functionality of excel can also be used to select the entire column.

* * *

Exploring some Excel Settings

The operation of some procedures in Excel Macros can be changed by one or more Excel's settings. One interesting setting is the Boolean setting. In this setting, you give a False or a True value to the excel procedure and then the setting is altered accordingly. Let's look at an example. In Excel Macros, you can give page breaks on your worksheet. You can turn on the Page break feature of excel by passing a real value to the DisplayPageBreaks procedure.

```
ActiveSheet.DisplayPageBreaks = True
```

Similarly, we can turn off this feature by passing a false value.

```
ActiveSheet.DisplayPageBreaks = False
```

Pacing up your VBA Code

VBA is a first way to perform data manipulations on your worksheet. In this section, we will be discussing some useful tips to further optimize the VBA Code to pace your operations.

One feature of excel is that whenever you execute the Excel Macros, the updates are visible on the screen. This, however, can have an impact on the performance of your code. You can, however, disable the setting. To do this, write down the below code.

```
Application.ScreenUpdating = False
```

This will turn off the screen updates while the Macros is being executed. To turn on the feature again, execute the below code.

```
Application.ScreenUpdating = True
```

One other way to speed up your Macros is by disabling the automatic calculation feature. If there are a lot of complicated formulas on your worksheet, by setting the calculation mode to the manual can significantly speed up things in your code. Execute the below statement to make the calculation manual.

```
Application.Calculation = xlCalculationManual
```

Macros display alerts messages to the users while the code is being executed. In this case, if the Excel is unattended and the Macros is executing then the alert message will bring the code to a halt. They require the humans to respond to the alert messages. However, there is an option to disable the alert messages so that the Macros is not brought to a halt.

```
Application.DisplayAlerts = False
```

You can also turn the alerts by writing a statement at the end of your code. To do this, we will just change the DisplayAlerts statements to a True condition.

```
Application.DisplayAlerts = True
```

Another to pace up the speed of your Macro Program is to ensure that you declare variables at the start of your code. It is imperative that you declare the data type of all the variables. Although

Excel doesn't throw an error if a variable's data type is left undefined but Excel won't know the exact size of the variable. As a result of this, Excel might assign space much more extensive than what it is required. This will result in extra memory consumption and can also decrease the performance of your code.

Chapter 4: Loops

Repeating blocks of VBA

Loops are essential as they make macros more capable and they also make the code easier to write. Instead of writing numerous statements for every cell on the worksheet, loops help to simplify your code. Several types of loops are supported in VBA.

For-Next loop in the programming language is referred to as the simplest type of loop. There is a control variable that acts as a counter to the loop condition. The counter begins from the start value and continues to be executed till the end value is reached. Code that is written between the For statement and the next statement is repeated in the loop. Let's look at an example.

```
Sub Multiply()

  Dim Product As Double

  Dim Counter As Integer

  Product = 0

  For Counter = 1 To 50

  Product = Product * Counter

  Next Counter

  MsgBox Product

End Sub
```

In the above example, we are performing a multiplication operation in the For-Next Loop. The counter will begin from 1 and continue to execute till the counter value is reached 50. There is the only statement written in the For-Next loop. The statement multiplies the counter value with the Product value and stores it in the Product variable. When the For-Next loop counter value is reached till 50, the Product value will be displayed on the screen. However, it is not advisable to change the counter value in the For-Next Statement as it can generate unpredictable results. For-Next loop can also include Exit statement within the block. The Exit statements are placed in the For-Next block to terminate the loop immediately.

In the following example, an Exit Statement has been inserted in the For-Next loop.

```
Sub Exit_Statement()

    Dim a As Integer

    a = 10

    For i = 0 To a

        MsgBox ("The value is i is : " & i)

        If i = 4 Then

            i = i * 10

            MsgBox ("The value is i is : " & i)

            Exit For

        End If

    Next

End Sub
```

The MsgBox will display the value of i. The value of i is incremented. As shown in the above example, when the value of I will be equal to 4, the code will enter in the if statement. In the if statement, the value of i is first multiplied by 10 and then it will be printed on the screen. Soon after this, an exit statement will be executed which will end the loop immediately. When the above code is executed, the following output will be displayed on the screen sequentially.

```
The value is i is : 0

The value is i is : 1

The value is i is : 2

The value is i is : 3

The value is i is : 40
```

The exit statements can be beneficial especially when we want to handle exceptions or errors in our code. If for example, there is a chance that a garbage value can come in a specific variable. I don't handle the garbage value, there is a chance that loop will be executed till infinity. This can result in memory overload. To cater to this scenario, the exit statements are placed so that the loop is terminated immediately.

Another type of loop is the Nested For-Next loop. The nested statement comes in handy when you want to loop through tabular data or multidimensional data. A table has data placed in both columns and rows. One loop is used to traverse through the columns whereas the other loop can be used to traverse through the rows. This is how you use the nested For-Next loop.In the following example, we will be filling up data in both the rows and the columns.

```
Sub Fill_Table()

  Dim Col As Long

  Dim Row As Long

  Dim i As Integer

  i = 0

  For Col = 1 To 3

  For Row = 1 To 3

  i = i + 1

  Cells(Row, Col) = i

  Next Row

  Next Col

End Sub
```

In this example, we will be filling up a 3 columns x 3 rows table. The outer for loop fills up the data in the columns whereas the inner for loop fills up the data in the rows. The output of the code can be seen below.

	A	B	C
1	1	4	7
2	2	5	8
3	3	6	9
4			

Another type of looping structure which is supported the VBA is the Do-While loop. The statements in the Do-While will keep on executing repeatedly until the mentioned conditioned is fulfilled.

```
Sub DoWhileDemo()

    Do While ActiveCell.Value <> Empty

    ActiveCell.Value = ActiveCell.Value * 2

    ActiveCell.Offset(1, 0).Select

    Loop

End Sub
```

In the above example, a Do-While loop is inserted. The Do statement checks whether the active cell is empty or not. If the cell is not empty, the active cell is multiplied by and then the next cell is select. Whenever the code will encounter an empty cell, the do-while loop will break. One should always ensure that there should be some break statement in the loop else the loop will keep on executing till infinity. This, however, can result in memory overflow. Loops considerably save up the space of the code and make it look compact and less complicated. However, loop statements do not reduce the time the code takes to execute.

Chapter 5: Arrays

Storing collection of elements

Arrays are usually supported by all programming languages. Arrays are also supported by VBA thus making life easier for programmers. An array is used where we want to collect data of similar type into one single variable. In an array, data is stored in a sequential manner. Each element in the array is provided an index number. The index number can be stated as a reference to the elements stored in the array. For example, we can define an array to store the names of the days of the week. If the array is named as Weekdays, the first element in the array will be stated as Weekdays(1). Similarly, the second element will be stated as Weekdays(2), the third as Weekdays(3) and so on.In order to use the array, it is mandatory to first declare the data type of array. VBA does not give us the flexibility to leave an array undeclared. Just like other regular variables, an array can be declared by either using a Dim statement or a Public statement. However, one additional thing about the array is that you need to mention the number of elements the array will hold. The syntax for declaring an array can be seen below.

```
Dim Dec_Array(0 To 10) As Integer
```

You declare the first index number, the keyword To and then finally the last index number. All this is enclosed inside the parentheses. One flexibility VBA offers is that you don't necessarily have to mention the lower index every time. In this case, VBA assumes that the lower index is 0. For example, the above statement can also be written as:

```
Dim Dec_Array(10) As Integer
```

VBA by default assumes the lower index to be zero. However, if you don't want VBA to assume the lower index to zero, you can use the Option Base statement to force VBA to change the lower index accordingly. For example, if you want VBA to assume the lower index to be always set to 1 then you should write:

```
Option Base 1
```

You also need to ensure that this statement is written before the declaration of the array.

So far, we have discussed one-dimensional arrays. A more natural explanation of the one-dimensional array is a single line of a value of the same data type. When we want to have more than one dimension in an array, this is known as the multi-dimensional array. VBA can handle up to 60 dimensions in the multi-dimensional array.

However, this is a sporadic case in which 60 dimensions are being used in the VBA code. A multidimensional array is also declared using the Dim statement.

```
Dim Multi_Dim(1 To 10, 1 To 100) As String
```

This is a multi-dimensional array. The first index of the multi-dimensional array ranges from 1 to 10, whereas the range for the second index is from 1 to 100. This results in the declaration of 1000 elements in the array. The numbers are stored in a tabular arrangement of 10 rows x 100 columns. To refer any element in the array, you have to mention two index number. The first index number will be the row and the second index number will be the column. For example, if we want to assign a value to the element in the 2nd row and 4th column, we will write it as:

```
Multi_Dim(2, 4) = 10
```

The value 10 will be stored in the element of the array that is placed in the 2nd row and 4th column. Dynamic arrays are also supported by VBA. The advantage about using a dynamic array is that you don't have to declare the number of elements it will hold at the start of your code. The array size is declared at runtime. Dynamic arrays help to save memory.

Chapter 6: Functions

Performing Specific Tasks

Let's first define what exactly is a function. A function can be called a procedure that performs some sort of calculation. A function returns a single value. Let's take an example, the Sum function will return the sum of the values that it will intake. Similarly, in VBA macros the function undergoes a specific calculation and then finally return a single value.

The function that you utilize in your VBA code primarily comes from three areas:

✓ Worksheet functions
✓ Some Built-in functions
✓ Customize functions that are defined according to

your needs

* * *

Playing with built-in functions

Well the good thing about using VBA is that it has made the life of programmer easy by defining some built-in Functions. You don't have to write every function from the scratch. You just need to know the exact name of the function and Voila! Let's look at some built-in functions of the VBA tool.

```
Sub DisplayDate()

MsgBox "The Date is: " & Date

End Sub
```

The above procedure has been created to display the system date. In this scenario, we have used the built-in excel function, Date. The Date function doesn't require any input arguments. We can also retrieve the time by writing Time instead of Date. Time again is also a built-in function of the excel tool.

Let's look at another example

```
Sub Get_String_Length()

Dim Name As String
Dim StringLength As Integer

Name = "Anum Haroon"
StringLength = Len(Name)

MsgBox Name & " has " & StringLength & " characters"


End Sub
```

In the above example, we wanted to retrieve the string length of the name. We, therefore, used the excel built-in function, Len().

If we want to retrieve the year or the month from the system date.

```
Sub Display_Month()

Dim Month_Name As Long

Month_Name = Month(Date)

MsgBox MonthName(Month_Name)


End Sub
```

In the above procedure, we are basically using the Built-in Month function to retrieve the Month from the System Date. Further, we are also using the MonthName Function in order to display the month name.

There are specific functions in the VBA tool that come up with some additional functionalities. For example, the MsgBox is a convenient function. Every time a user uses the MsgBox function in its Macros, a screen is popped up at whatever line it is executed. This MsgBox can be very handy to prompt the user about the values that are stored in the variables. Moreover, it also helps to debug the code and find out any potential errors in the code. There is another convenient function which is known as the InputBox function. The InputBox function gives the user the capability to enter a value into a simple box that is displayed. The value that is retrieved through the InputBox can be further manipulated in the macros that we have written.

The VBA tool facilitates the user with lots of built-in functionality. The question is that how do you locate those functions? Well, that's not an issue at all. You can retrieve a list of all the built-in function by typing VBA followed by a period as shown in the diagram below

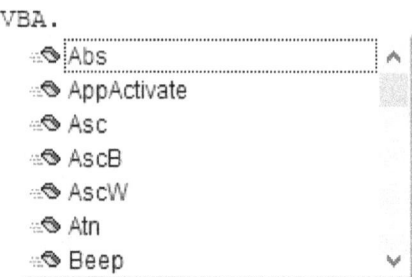

In order to find out the details linked to a particular function, enter the name of the function in the VBA module, move the pointer on the text and then press F1. You will be prompted with a new window screen and help wizard will open up.

Although the list of inbuilt functions is exceptionally long, I have compiled some important ones in the table below.

Exploring the worksheet functions

One other interesting feature of the Excel VBA is the worksheet functions. A worksheet is an area where the cells are placed and the data is manipulated on the cells. In a nutshell, the worksheet is the workspace where basically all the calculations and the manipulations take place. When we are accessing functions using the "WorksheetFunction" expression, the manipulation or calculation is performed on the active sheet. By active sheet, I basically mean the sheet that is currently open on the screen.

Let's look at an example of a worksheet's sum function.

```
Total_Amount = Application.WorksheetFunction.Sum(Range("B1:B3"))
```

In the above example, "Total_Amount" is the variable in which the sum of the cells is being stored. Range("B1: B3") refers to the cells on the worksheet of which the sum has to be taken. VBA also gives the user's to access the Sum function directly either from the Application part of the WorksheetFunction part. VBA has the ability to figure out what exactly you are performing. Smart Right? The following three statements have exactly the same output.

```
Total_Amount = Application.WorksheetFunction.Sum(Range("B1:B3"))

Total_Amount = WorksheetFunction.Sum(Range("B1:B3"))

Total_Amount = Application.Sum(Range("B1:B3"))
```

I personally prefer to perform my calculations using the WorksheetFunction to have a better understanding of the code that is executing. Let's look at some more Worksheet Functions.

To calculate the Min Value in a range

```
Sub Find_Min()

Dim MinNum As Double

MinNum = WorksheetFunction.Min(Range("B1:B10"))

MsgBox ("The Minimum Value is " & MinNum)

End Sub
```

In the above example, the Minimum value is calculated from the range of cells from B1 till B10. The value is stored in the variable MinNum. The output of the above expression can be seen in the diagram below

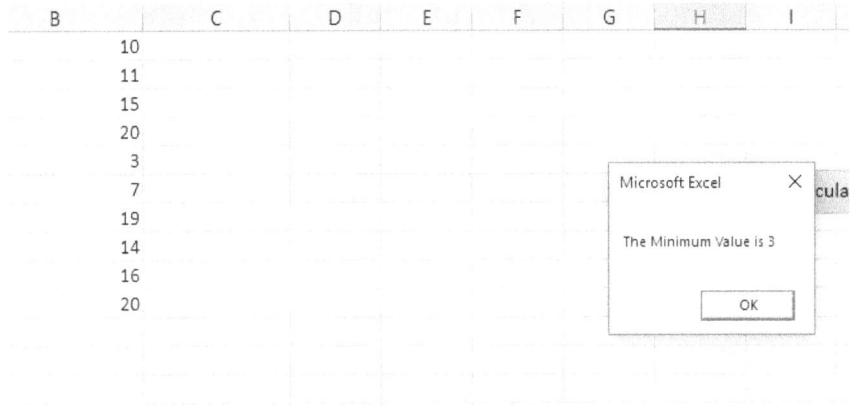

Similarly, you can also calculate the maximum value from the Range using the Max() function of the worksheet.

Vlookup Function

If you have ever interacted with an excel programmer, I am sure you must have heard about the Vlookup Function. You can never be an Excel Guru if you don't know how to implement the Vlookup Function. Vlookup is a potent function as it can help you the exact information from a table of any size. The Vlookup Function intakes four parameters. The first parameter represents the item that needs to be searched in the table. For example, in the below example we are looking for the product B. The second Parameters represents the range in which it needs to be looked up, the third parameter represents the column number that we want to return as a result, and finally, the four parameter intakes a true or a false value. The true value tells the vlookup to return a value that can be an approximate match whereas the false value tells the vlookup to only return a value when there is an exact match. In the following example, the user will be entering the product name in order to retrieve its price from the table. The product name will be entered using the InputBox Function. The price of the product will then be displayed on the output screen using the MsgBox Function.

```
Sub Get_Product_Price()

Dim Product

Dim Price As Integer

Product = InputBox("Enter the Product ")

Price = WorksheetFunction.VLookup(Product, Range("A2:B5"), 2, 0)

MsgBox ("The Product Price of " & Product & " is " & Price)

End Sub
```

As shown in the below diagram, the input box is prompted on the screen when the Macros is executed.

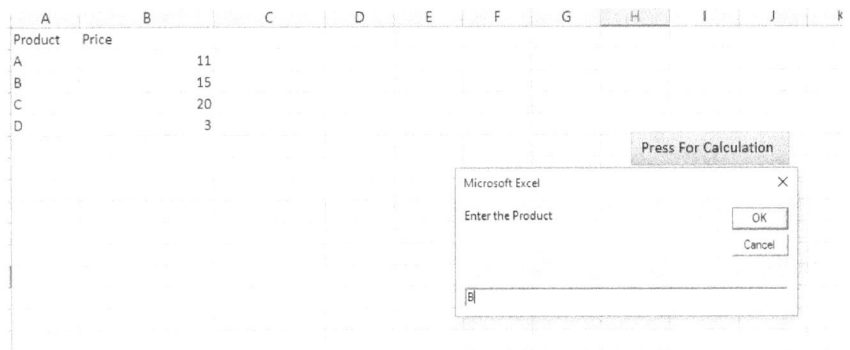

In the example, the Product B was Entered. On pressing the Ok button, the Output was displayed.

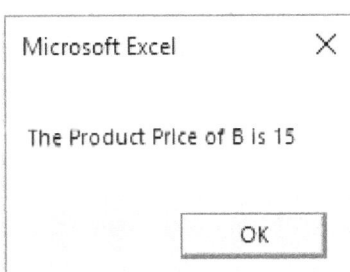

The Input was passed to the Vlookup Function. The Vlookup Function traversed on the Product B and returned the Price of the Product.

Conclusion

By the end of this chapter, you must have become very much familiar with the Excel VBA. In this section, I will just briefly gives some essential guidelines.

It is always a good practice to declare your variables at the start of your code. Leaving your variables undefined will only save you from few seconds but later on, you will have to bear with the consequences. So make a habit to declare all the variables at the start. This will significantly impact on your code quality.

Once you are done finalizing your code and everything is running perfecting, make sure you do the last minute cleanup of your code. Make sure all the code lines are indented. This will help you understand your code better when you will review it later. Make sure all your variables are declared and the description of the variable should be relevant to their actual operation. For example, if a variable will be holding Amount in it so a good practice it is to name it as "Amount" rather than naming it as XYZ. It is always a good practice to add comments in your code. Do a quick check to remove any redundancy in your code.

As Macros can compromise multiple procedures, therefore you shouldn't avoid putting all the calculations in one procedure. For example, the procedure for calculating Sum should be in a separate procedure. The procedure for converting string values to date value should be in a separate procedure and so on.

One last advice is to make sure that you take a backup of your code every now and then. It takes an effort to make a correct macro code and there is always a chance that your Excel Workbook can get corrupted. So to save yourself from all the re-work, make sure to take a backup after every few days.

EXCEL

MACROS

Programming For Complete Beginners, Step-By-Step Illustrated Guide to Master Excel Macros

Introduction

The aim of this book is to provide complete information about the Macros in excel. Starting from the very basic explanation of Macros, we will dig deeper, step by step in each chapter briefly detailing all the important properties of Macros and what you need to know to get started in this domain.

A very disciplined approach is taken for writing this book which will be very easy for readers to understand, who we assume are very new to this topic. We will explore all the areas relating to Macros and will also go in to the VBA(Visual Basic for applications). We will also mention some important techniques for recording Macros and how Macros are actually stored in excel(Which format?).

After reading this book, the reader will have complete knowledge about Macros, VBA(Visual Basic for Applications), the relationship between VBA and Macros, what is the difference between Absolute Macro recording and Relative Macro Recording, what are VBA Excel Objects, how to send an e-mail directly from Microsoft excel and a dedicated chapter defining a complete approach for debugging. Apart from this many other small level details will also be explained, so that our reader doesn't feel the need to google anything and deviate from the actual topic.

Chapter 1: What are Macros?

Macros

A macro is basically a program/action which is a replacement for recurring mouse or keyboard actions in Microsoft Excel. Macros are used to help save the time of users by automating some repetitive tasks that you have to do over and over again. In Excel, creating a Macro which is also termed as recording a macro is done using Macro recorder. The recorded Macros are written in Visual basic for Applications(VBA). For now, you only need to know that VBA is a programming language which is used for creating Macros. Since Excel is a Microsoft product, VBA is also developed by Microsoft.

More brief definition of a Macro

Now you have a basic idea about Macros. You know that they are written in VBA, but how can a person specifically define a Macro to a professional person. For example, you cannot say that Macro is a VBA. We all know Macros are written in VBA but professionally the correct way to define a Macro to a technical person or programmer is: "Macros are Visual Basic procedures that are used to automate tasks in Excel, saving users time and efforts".

Macro Recorder

One way of creating Macros is by writing instructions or by coding directly in VBA editor. This approach is used by developers who are properly aware of Macros and are familiar with coding to a great extent.

Another approach is used to generate VBA code automatically. This approach is useful for novice users who are not familiar about coding or syntax of the VBA. This approach is achieved through Macro Recorder. As you interact with data in the excel, the Macro recorder examines your actions and them automatically generates VBA code respectively.

For further convenience, we can link the macros to different keyboard shortcuts. In this way, the keyboard shortcuts will act as commands for triggering a Macro.

Is Macro Recorder enough for all our automation needs?

Now, one might ask, if you have Macro Recorder, what is the purpose of manually giving instructions for a Macro in a VBA editor? The answer to this is very simple but is of immense importance if you want to know about significance of coding in Macros: There are certain functionalities which Macro Recorder cannot record or provide automatically. For example, Macro Recorder cannot provide loop functions and screen prompts. To achieve this functionality, you need to code in VBA editor.

Chapter 2: Getting Started with VBA

VBA (Visual Basic for Applications)

VBA is a programming language that is developed by Microsoft, in Excel the Macros are programmed in this language. So, it is very important for you to know about VBA because you will be spending most of the time coding Macros in this language.

Now before moving further, you should know that the VBA is not limited or restricted to Microsoft Excel only, it is also very consistently used with applications like MS-word and MS-Access(Microsoft's own database Application).

What can you do with VBA?

Some tasks require complex calculations which cannot be satisfied with in-built functions of Excel. This is where VBA plays its part, you can write set of instructions for performing calculations and create Macros comprising of these instructions. You can use these Macros again and again as long as it satisfies your needs. This would not be possible without the ability to build customize Macros.

Creating custom commands, new spreadsheet functions and functions for performing repetitive and frequent actions, all of this can be done by coding Macros using VBA.

Using the VBA

As mentioned before, all of the coding is done in VBA editor. To open VBA editor from Excel, just press 'Alt+F11' and a new window will open as shown in the figure below:

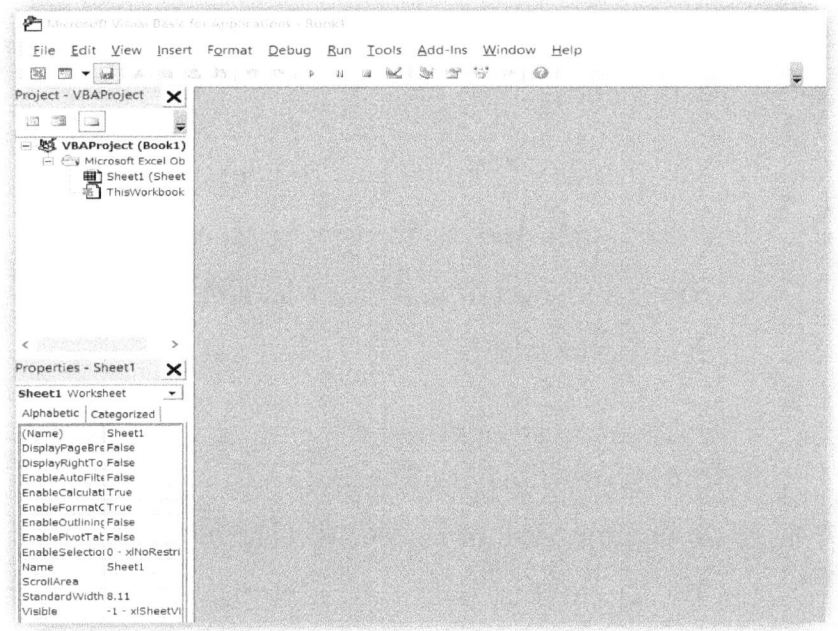

This is where you will write your Macros. When inside VBA editor, click on 'insert'. Few options will be displayed in a drop-down list. From the drop-down list select the 'Module' option. This will open a new window. In this window, you will write your

Macro commands. Don't worry, we don't expect that you will write commands yet. We just wanted to show you the place in the editor where we will be writing our Macro commands.

Now, since you know how to open VBA editor and 'Module' for typing Macro commands, its time that we learn about the developer option.

The developer option
Before we start programming in VBA, we need to do one more step: Enable the Developer option. To do this, follow these steps:

1. Open a workbook in Excel.

2. Right click on the ribbon inside Excel workbook.

3. Click on customize ribbon.

4. Check the developer option.

As you can see in the picture given below, we have checked the developer option.

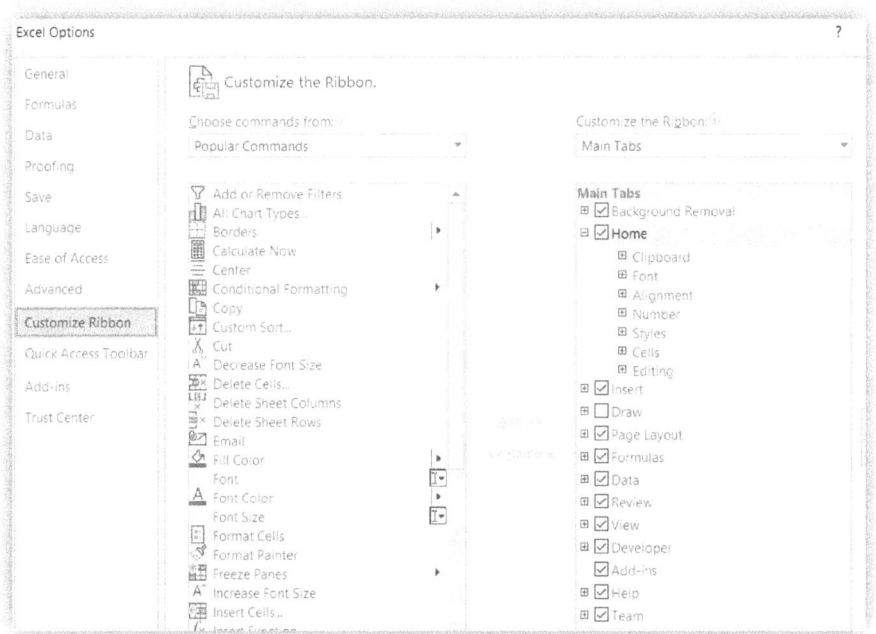

Shortcut for opening VBA editor

Earlier we showed you a shortcut for opening VBA editor, a different way, which takes couple of more steps, is to click on the 'developer' in the workbook ribbon, then click on the Visual Basic button, resulting in the opening of new window in VBA editor.

(Shortcut: Alt+F11)

Examples of function procedures
Here we shall look at the method of writing function procedures. Function procedures are very helpful in VBA code as they can be written to return a value and can be called as many times as you like.

To write a function we use the reserved keyword "Function" followed by a user-given name. An argument can also be passed to the function. An argument is basically a parameter with which we can interact with inside the function body to generate some useful result.

For example:

Function sum(a, b)

sum = a+b

End function

The function in the above code takes two parameters 'a' and 'b', adds them and stores the result in a variable called Sum.

With the above example, you must have had an idea about importance of function parameters. You can pass as much as **255** arguments to a function or you can pass non. That's right, you can even write a function without giving an argument.

Example of sub-procedure

Referring to an object inside the sub-procedure

Excel VBA contains a very interesting function called "HasFormula". This "HasFormula" lets you know that whether a specific cell inside the excel workbook contains a formula or not. If the cell contains a formula, it returns true, otherwise it returns false.

Sub check ()
Dim checkFormula **As** Boolean
checkFormula =
Range("B1:B2").**HasFormula**

MsgBox checkFormula

End Sub

In the above check() sub procedure, first of all we defined a variable with 'checkFormula' type 'Boolean' and then used '.HasFormula' with the Range("B1:B2") which returns true or false depending on whether or not the cells **B1** and **B2** contain a value or not. MsgBox is another function which is used for displaying values.

Referring to property of the object 'Columns'

We shall study about Excel objects in detail in the chapter 6. We shall write a function that sets the value of a specific range to "", i.e. It clears the contents of a particular range.

Function clearRange()

Columns.("A:A").clear

End **Function**

The second line in the above function clear the contents of the cell : "A".

Chapter 3: Macro Security

At this stage, you have a basic concept about Macros and VBA. In this book, we will be moving very slowly towards more advanced topics, addressing all the aspects of every single topic.

Macro security is very easy yet very sensitive topic. The reason it is considered sensitive is because if not understood properly it can be dangerous for your computer because without proper security you can get viruses through Macros, which ultimately can take the control of your whole computer and can harm your privacy.

Security settings

Inside Excel workbook under the 'developer', there is an option called 'Macro Security'. Click on the 'Macro Security, and a new window will appear with a list of options. The name of this window is Trust Center, highlighted as shown below.

Trusted Locations

In 'Trust Locations', all of your trusted locations for opening files will be listed. If you want to add a new location, always make sure that the new location is trustable and secure. Otherwise you can get dangerous malware and viruses from unknown sources. Never add unknown location for opening files, if you want to have a secure and computer-friendly environment (who doesn't want it).

Storing macros in your Personal Macro Workbook

To store the Macros in your personal Macro Workbook, first you need to create a Macro. To create a Macro go to 'Developer' tab. Under the 'Developer' tab click on 'Record Macro'. You will be asked to type in the name for your Macro to save in the workbook as shown in the figure below.

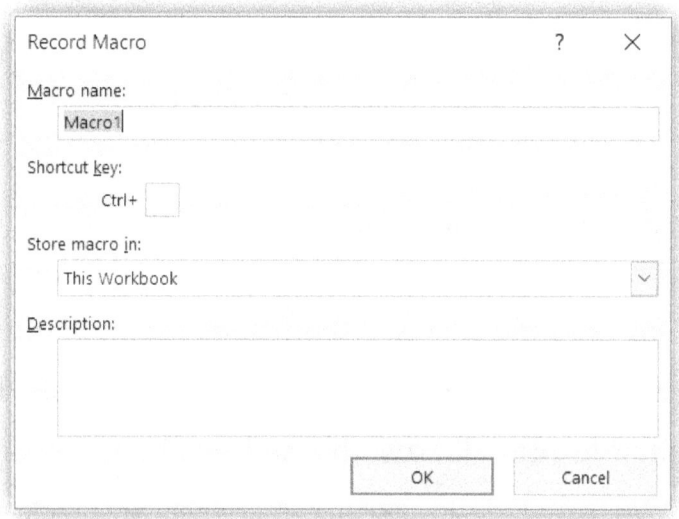

You will see a text field as shown in the figure above that says, "This workbook". Since we want to store the macro in our personal workbook, we will click on the list for it to drop down. Then click on Personal Macro Workbook and then press 'OK'. Now as the name is set we shall perform actions which we want to store in the Personal Macro Workbook. After that click on Stop Recording under the 'Developer' tab as shown in the figure given below.

☐ Stop Recording

⊞ Use Relative References

⚠ Macro Security

Now close this workbook and you will be asked if you want to save the changes. Click on 'save'. You have successfully stored Macro to your Personal Workbook.

Macro settings

Another very important option inside the trusted window is 'Macro settings'. This option gives you variety of choices as shown below in the figure:

Macro Settings

- Disable all macros without notification
- Disable all macros with notification
- Disable all macros except digitally signed macros
- Enable all macros (not recommended; potentially dangerous code can run)

Developer Macro Settings

- Trust access to the VBA project object model

Now we will explain the options in Macro settings:

Disable all Macros with notification

With this option Macros will be disabled, but you will be alerted if there are any security notifications regarding Macros that are present.

Disable all Macros without notification

All the security notifications and all the Macros are disabled with this option

Disable all Macros except digitally signed Macros

Macros are disabled, but if the Macros is signed by a trusted published then it will run.

Enable all Macros (not recommended)

With this option you can run all types of Macros whether trusted or not. This setting is usually not recommended as it makes our machine vulnerable to viruses and malware software.

Chapter 4: Recording your first Macro

If you have understood the VBA then this is just piece of cake for you. Recording the Macro doesn't mean that you will write the code in VBA and it will show the output in Excel workbook. The latter is true that the output will be showed in the Excel workbook, but prior is partially true. We said partially because ultimately VBA code is responsible for Macro behavior, but you will not write it. It will be automatically written.

Steps to record your first Macro
Now before we start, we would like to mention something here, the steps we follow will create a Macro which will do a task we want Excel to do. You can follow any steps you want as long

as you understand the concept(how to record and use Macro).

1. Go to 'developers' tab.

2. Click on 'Record Macro'

3. When the Macro has been clicked, any action you perform will be monitored by Excel and will be converted to VBA code.

4. In this example, we will be determining Annual pay of person. We will use a formula called 'product' which you all are aware of: It multiplies two quantities.

5. While the Macro is recording, we will perform steps as shown in the figure given below.

	A	B	C	D	E
1	Monthly pay				
2	20000				
3	Annual Pay				
4	240000				
5					
6					
7					

6. After we have performed these steps, we will stop recording and save our macro.

7. Create a shortcut for your Macro. The shortcut will be of your choice, so you can choose any shortcut. For this example, we will be choosing 'Ctrl+c' as our shortcut key for calling our Macro.

Running a Macro

Once we have saved the Macro, now we can call the Macro using our shortcut key. As soon as we shall press 'Ctrl+c', all the steps you did while recording Macro will be repeated and you will see the output on the Excel sheet.

We mentioned earlier that all of the moves that you do while recording Macro are converted to VBA code. So, where is this VBA code? This VBA code is stored in Module.

As shown below, following are the lines that were generated while I was performing steps during Macro Recording.

```
Book1 - Module1 (Code)

(General)                                                    Calculate_yearly

Sub Calculate_yearly()
'
' Calculate_yearly Macro
'
' Keyboard Shortcut: Ctrl+c
'
    Range("A1").Select
    ActiveCell.FormulaR1C1 = "Monthly pay"
    Range("A2").Select
    ActiveCell.FormulaR1C1 = "20000"
    Range("A3").Select
    ActiveCell.FormulaR1C1 = "Annual Pay"
    Range("A4").Select
    ActiveCell.FormulaR1C1 = "=PRODUCT(R[-2]C,12)"
End Sub
```

The above window shows the operations performed during Recording converted to VBA commands. Two words are highlighted, these show the name of the

Macro. Active Cell refers the current cell where pointer is pointing.

Assigning a macro to a button

Follow these steps to create your first Macro and assign it to a button:

1. Inside the 'developer' Tab, click on insert.

2. Click on the button as shown below.

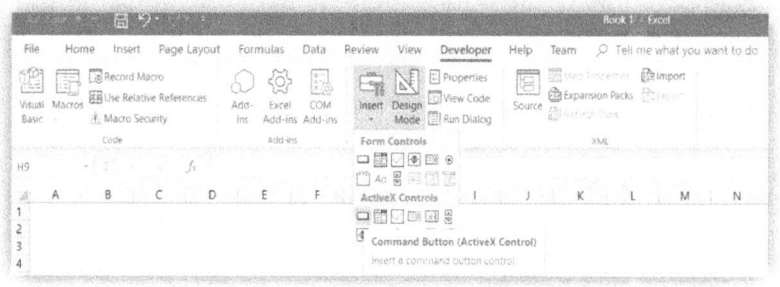

3. Then left-click once, then should be able to see a button as follows:

4. Right-click on by keeping the mouse arrow on the button.

5. Select properties.

6. Select the name of the button and its caption. What you write in the caption will appear on the button.

7. As an example, here is the name and caption of my button.

Properties		
Button1 CommandButton		
Alphabetic	Categorized	
(Name)	Button1	
Accelerator		
AutoLoad	False	
AutoSize	False	
BackColor	☐ &H8000000F&	
BackStyle	1 - fmBackStyleOpaque	
Caption	Click on this button	
Enabled	True	
Font	Calibri	
ForeColor	■ &H80000012&	

8. Now left-click on the button twice and a new window will open in VBA editor. This is where you can code yourself in the sub-procedure which will take the commands. For example, you can write the code that what happens when you click the button. This is just one of the possibilities, you can make it perform several other tasks as well. Again, it all depends on you, you can write code according to your needs.

9. Add any command, for your learning, I have added a simple command which will show a message "This button is clicked",

whenever anyone clicks the button.

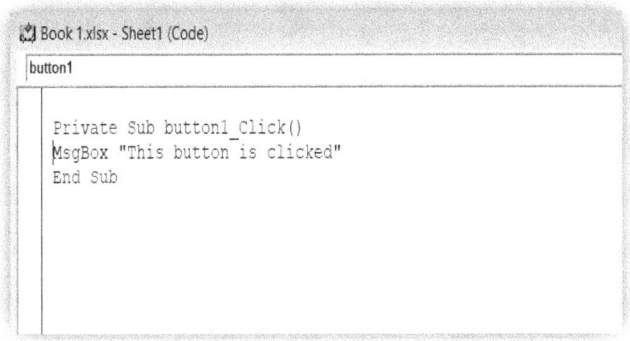

```
Book 1.xlsx - Sheet1 (Code)

button1

Private Sub button1_Click()
MsgBox "This button is clicked"
End Sub
```

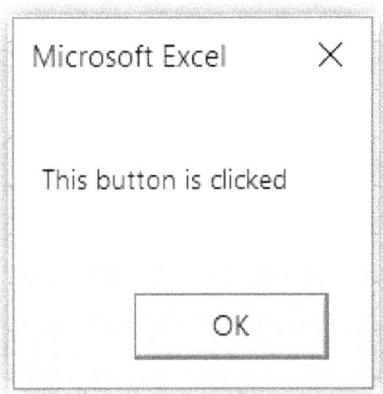

```
Microsoft Excel         ✕

This button is clicked



        OK
```

10. Now if you have created a button and want to delete, whatever the reason, you can do so easily by clicking on the 'design mode' under the 'developer' and then selecting the

button and pressing the following keyboard button "delete". There you go, button deleted.

After following all these steps, you have successfully created your first Macro. It was easy, wasn't it? If you keep following all the steps systematically and keep learning patiently then you will be able to master all Macro techniques easily.

Saving a Macro-Enabled Work Book
When we work with Macros inside the Excel, we need to save the workbook in a Macro-enabled format. The reason for this is that it provides an added security from external threats. To do that first we have to click on 'file' tab on the top left of the screen. Now if you move down some options you will see an option that says, "save as". Click on it and then you will be asked to select the location for saving the macro. Choose any location according to your need or priority. Once you

have selected a location, click on the drop-down list as shown in figure below.

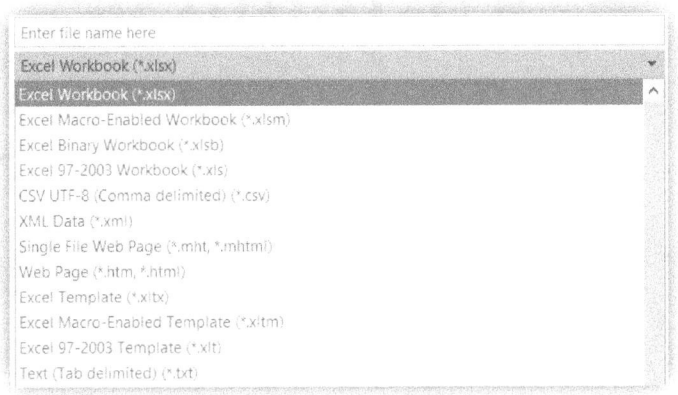

Click on the format that says, 'Excel Macro-Enabled Workbook(*.xlsm)'. You have successfully saved a Macro enabled file.

Chapter 5: Absolute vs. Relative Macro Recording

If you performed the Macro Recording activity in the "Recording your first Macro", you may have noticed that no matter how many time you run the Macro it always occupied the first four rows. It didn't matter where you pointed in the Excel worksheet, no matter which cell you are pointing, whenever we ran the Macro, it filled first four rows. Now, this is approach might not be effective as in worksheet you have to deal with large amounts of data and therefore, you cannot keep the number of rows fixed. You may have to work on 1000's of rows or may be even more. We need something that can put data relative to previous data. This is where the concept of Relative and Absolute Macro recording steps in. Here we will briefly explain both approaches and guide you which approach is the best one.

Absolute Macro Recording

In absolute Macro Recording, actual references of the cells are recorded. This means whenever you will run the Macro with absolute references it will affect the fixed references only. Those fixed references are the references provided while you were recording Macro. Affection of the fixed reference means actual cells are affected every time whenever you run Macro.

Relative Macro Recording

In Relative Macro Recording, it doesn't matter which cell references you interact with during the Macro Recording. When you will run the Macro after saving it, it will affect those cells which you are pointing to currently or it will affect cells relatively from original cells.

Which approach is best?

Without a doubt Relative Macro recording is feasible for us because absolute Macro recording is of no use as it keeps updating the same cells again and again. Relative in the other hand affects cells relatively. It doesn't keep updating or writing the same cells again and again. So, the choice is easy, Relative Macro Recording is more helpful compare to Absolute Macro Recording.

How to perform Relative Macro Recording?

For this purpose, you just have to perform one step before starting Macro recording. Click on

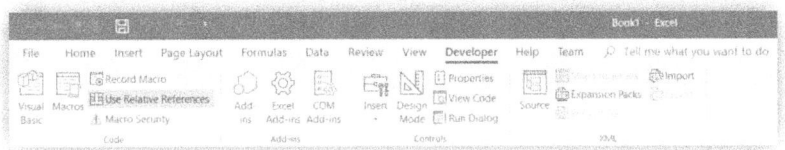

the 'Use Relative Reference' under the 'developer' tab. When you will run Macro after saving it, it won't update the same cells again and again. If you don't use the relative approach, then for absolute recording, you don't have to do anything, it automatically does absolute Macro recording. Check out the figure below for further assistance in Relative Macro recording:

Example of Relative Macro recording

The beauty of relative Macro recording is that you can run the Macro anywhere in the sheet, it will execute and show the results in the cells that you have selected(currently).

Here is an example in which we create a Macro in the cells: F6, G6 and H6 and generate the result in some other row or specifically in three cells other than these.

Steps:

1. Go to 'Developer' tab.

2. Under 'Developer' tab, click on 'Use Relative References'.

3. Click on 'Record Macro'.

4. When you click on 'Record Macro', a new window will open as shown in the figure below that will ask you about Macro name and where you want to store the Macro.

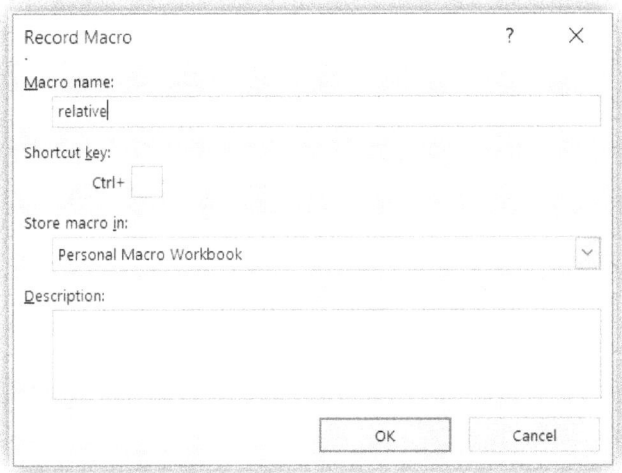

5. Now click on any cells you wish to update and write something in them.

6. We chose the cells 'F6, G6, H6' and wrote something as shown in figure below:

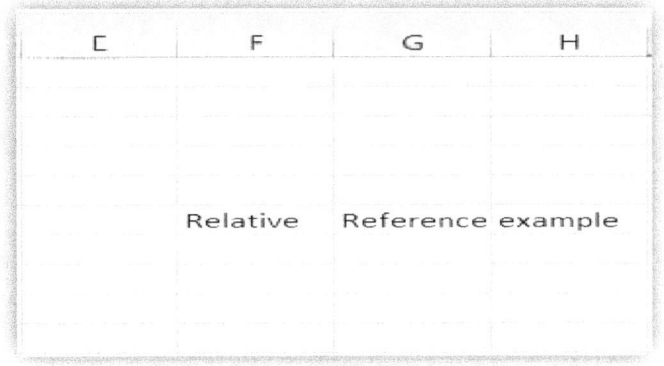

7. After you are done updating cells, click on stop recording. You have successfully saved a relative Macro.

8. Again under 'Developer' tab, click on Macros. A window will open as shown below. Click on run Macro.

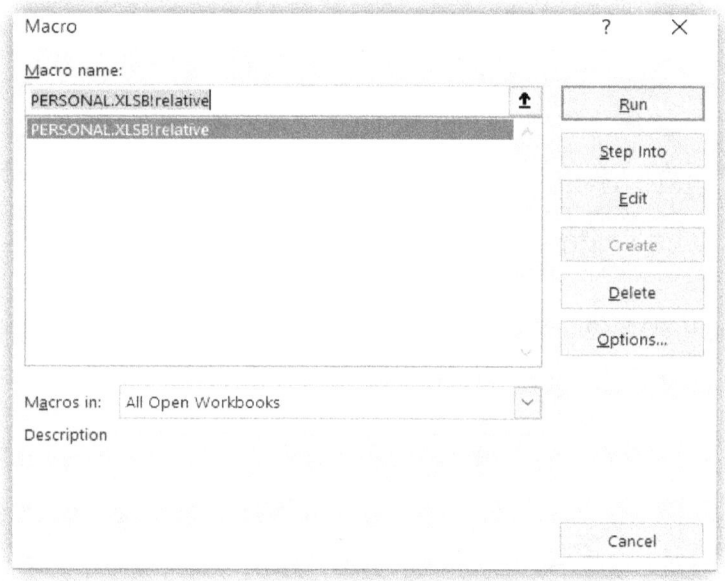

9. It doesn't matter which cell you have selected; three cells will be updated in row with the cell you have selected.

Chapter 6: VBA Excel Objects

If you are familiar with programming languages like Java and C++ then these topics will be very easy for you. VBA Excel object is used for the same purpose for which we use object in Java or C++. So, learning VBA Excel objects for you wouldn't be difficult. The only thing you need to learn is the syntax for calling the objects. As for my readers who are novice to this topic, you don't have to worry as I will be explaining everything briefly relating to objects and their purpose. So, keep on reading to grasp this concept like an expert.

Excel objects

The reason for all the explanation was to help you understand Excel objects properly in an effortless manner. As we said, objects perform some tasks. There are some objects inside Excel that perform tasks that are very useful to us.

Excel is an object too!

As mentioned above objects perform tasks for you. Here, the whole Excel application is an object too. It performs variety of tasks for us. Excel performs tasks like creating graphs or charts, organizing data, providing the ability to users to access Excel file from a range of different device, from different locations!

We also mentioned that an object can contain other objects. Same is the case for Excel. An Excel is an object which contain other objects. We shall present a list four objects that are very popular in Excel. We shall also be explaining in these concepts.

Following is the hierarchy of objects in Excel that a user will deal with, during most of his/her time spent on Excel:

1. Application Objects
2. Workbook Objects
3. Worksheet Objects
4. Range Objects

   (https://www.tutorialspoint.com/vba/vba_excel_objects.htm)

All of the objects are under the Application object. To call an object that is under Application object, you have to use '.'. You cannot call an object directly using Application object, you can call it with the reference of previous object which has to be called too.

- To get to a workbook that is named 'workbook' you have to write the following line:

Application.Workbooks("workbook.x lsx")

- Similarly, to open a specific worksheet in a workbook write the following line:

Application.Workbooks("workbook.x lsx").Worksheets(1)

The 1 inside the Worksheets parenthesis is used when there is only one worksheet in a workbook.

- Now suppose you want to retrieve a value from a cell named "B3", then you have to move further down the hierarchy to get this value. You need to write the following line in VBA to access this value:

Application.Workbooks("workbook1. xlsx").Worksheets(1).Range("A1").Va lue

Example of uses of an object

As mentioned earlier, each object has its own methods and properties. The methods are basically the functions which an object performs. Here we will give you an example on how to create a new worksheet in Excel using the object "Worksheets".

Let's create a new worksheet with name : "Hello"

Once inside VBA editor press "Ctrl+G", it will open up an immediate window in which you can start writing VBA code immediately. Write following line:

Worksheets.Add().Name = "Hello"

Now press enter. Go to back to Excel and you will see a new worksheet created under **current** workbook as shown in the screenshot given below

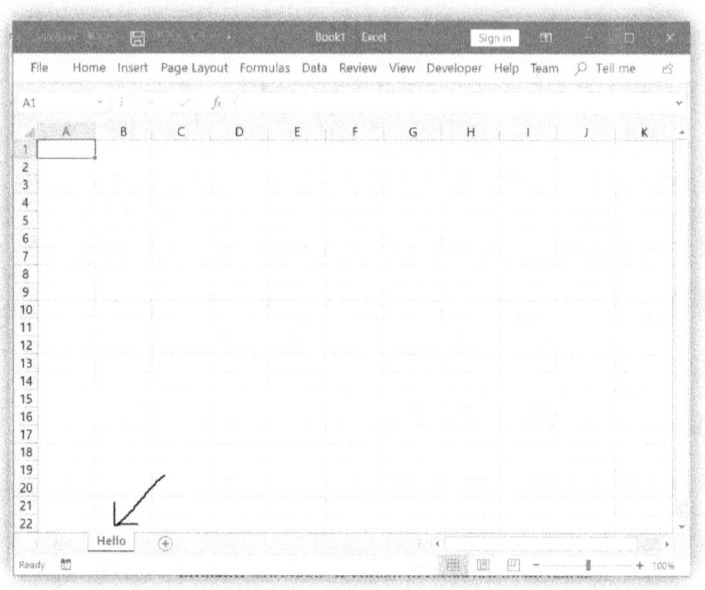

Properties of objects

Every object performs some tasks and each object consists of qualities or attributes that describe what the object is. For example, a car is an object and its properties are engine, color, model etc.

Similarly, VBA objects have properties which describe them. With the help of VBA we can modify the properties of settings. Confusing? This example will clear your concept.

The 'worksheets' object has property 'range' and the range object has the property value, written as:

Worksheets("Book1").Range("A").Value()

Now, value is a property of Range and we can play with its characteristics. You can set the value of a cell using this property and you can display this value in a message box using the function 'MsgBox'.

To display the value, you can write the following code:

Sub displayValue()

      Value = Worksheets("book1").Range("A").Value

      MsgBox Value

End **Sub**

Similarly, to set a value of a cell you can write the following code:

Sub **setValue()**

    Worksheets("book1").Range("A").Value = "Set Value"

End **Sub**

In the above example, we used the property value of the Range in two ways:

1. To display the value

2. To set the value of the cell.

Chapter 7: Sending Email from Excel

The topic we are going to discuss in this chapter can be very useful commercially if understood and applied in real-life situations properly. Every one of you must have used email sending and receiving platforms like Yahoo, G-mail, Hotmail, Outlook etc. But, have you used Excel to send email to someone or have you sent bulk of emails to same person at the same time? We assume your answer is :'no'. Well, in this chapter you will be expert in sending bulk of emails to multiple people or even one person.

Again, we will be using Visual Basic to achieve this task. If you have read this book from the very start, by this time you must be completely aware of how important VBA programming in Excel is.

Steps for sending the email

First of all go to 'Developers' tab after opening Excel workbook. Click on Visual Basic. A new window will open. Click on 'Tools' tab inside VBA. Click on references. A new window will be displayed on your screen containing bulks of libraries. Don't be overwhelmed by a large number of libraries. We will use only one of it.

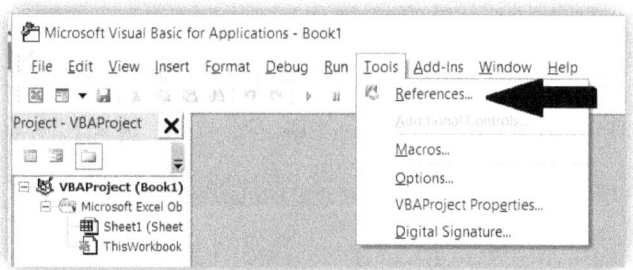

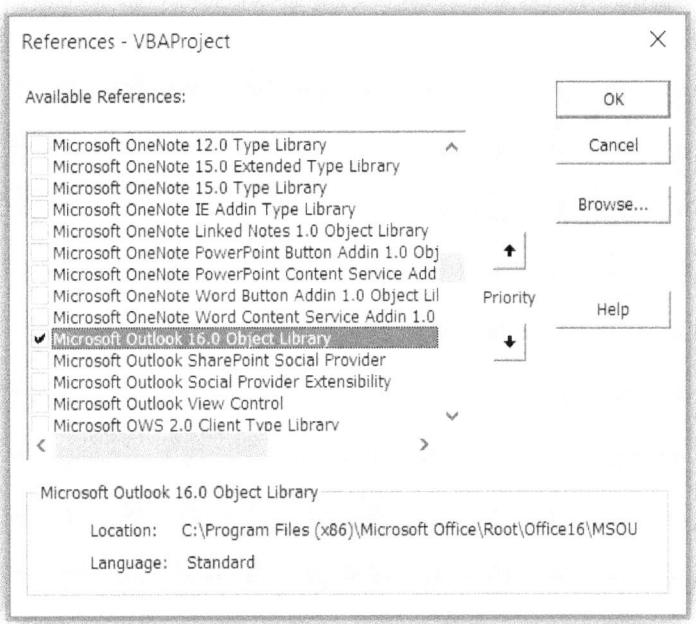

Scroll down the 'References-VBA project' window. You will see a library called Microsoft outlook 16.0(Version as of now) Object Library. Check this Library and click 'Ok'.

Now its time we actually start coding for sending an email to some person/organization.

To do it first of all click on 'Insert' in VBA window. In the drop-down list click on

'Module'. You have successfully created a module. Now it's time for writing some code!

Inside the Module, type in following code.

Sub SendMail()

Dim olApp As Outlook.Application

Dim olMail As Outlook.MailItem

Set olApp = New Outlook.Application

Set olMail = olApp.CreateItem(olMailItem)

With olMail

    .To = 'abc@gmail.com'

    .Subject = "Message from Excel"

    .Body = "Body of the message"

    .Display

```
    ' 'olMail.Send
```

End With

End Sub

When you write this code inside VBA and hit run, Excel will take you through a series of steps for configuring the email from which you want to send the email. After that email will be sent to the specific email you typed inside VBA code in front of '.To'.

So, it was easy sending an email through VBA? Sending email to many users is also easy you just add two more lines of code in the above code and Taddaa! Its done.

We need to add a 'loop' to the code. For the readers from 'non-programming' background, loop is just a statement in a code which allows us to execute a specific section of code repeatedly as many time as we like.

Coming back to our topic of sending bulk emails, following changing in the above code can help you send emails to multiple receivers(99 in our case):

Sub SendMail()

Dim olApp As Outlook.Application

Dim olMail As Outlook.MailItem

for i=2 To 100

Set olApp = New Outlook.Application

Set olMail = olApp.CreateItem(olMailItem)

With olMail

    .To = **Cells(i,1).Value**

    .Subject = **Cells(i,2).Value**

    .Body = **Cells(i,3).Value**

```
.Display

    ' '.Send

End With

End Sub
```

In the above code we applied **Bold** to some of the text to let you know about the changing or difference between initial code and this code.

Chapter 8: Debugging

As with all the chapters before, we shall begin this chapter by discussing a couple of terminologies. We shall discuss debugging Excel later in this chapter in detail. First of we should know what is debugging and how its effects our coding or to be exact: our computer program. Mind you, the concept of debugging is not limited to Excel. It is applied to almost every program in the computer. Why do we need it? Well, bear with me through this chapter and you will grasp the concept of 'Debugging'.

Debugging

Debugging is the process of removing "bugs" from computer programs that are the reasons for deviation of computer programs from expected behavior.

Bugs
Definition from Wikipedia:

"A software bug is an error, flaw, failure or fault in a computer program or system that causes it to produce an incorrect or unexpected result, or to behave in unintended ways. The process of fixing bugs is termed "debugging" and often uses formal techniques or tools to pinpoint bugs, and since the 1950s, some computer systems have been designed to also deter, detect or auto-correct various computer bugs during operations."

The importance of debugging is obvious from the definitions and explanations of both the "Bugs" and "Debugging". Now, let's jump straight into "debugging in Excel".

Debugging in Excel
The debugger tools in Excel allow us to pause program execution at any stage and then check

the value of variables and the continue executing the program. Exciting, Isn't it?

This can be done by using a breakpoint in the program. Setting a breakpoint is very easy and doesn't require any rocket science. All you have to do is to move the cursor to the statement where you want to pause the program's execution and press F9.

Here is an example of simple breakpoint in program:

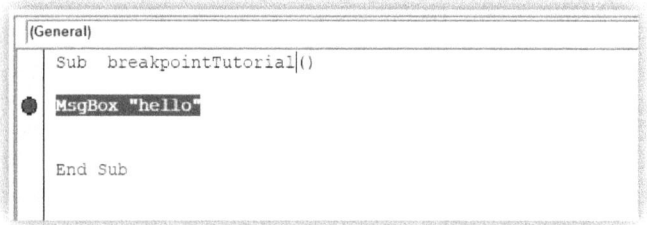

The red circle at third line shows that a breakpoint has been set here. When you run the program, the VBA editor halts program execution when it reaches breakpoint. After you have checked your program, to continue the execution, click on small green 'play' button to complete the program's execution and get normal output as always.

You can set the breakpoint of the program in another way, you can write one word in the VBA editor Module to halt programs execution. The word is: 'Stop'. Write this word anywhere in the program where you want to set the breakpoint. When the sequential execution of program will reach this statement, our program will enter into debugging code.

Moving down the code, statement by statement

When you set a breakpoint, the program halts its execution at the place where it is set. Suppose you want to move through some section of code step by step. This can be done by pressing F8. Each time you press F8, you move to next step. This is very useful and fast way of debugging rather than setting breakpoints at many points in the program!

Conclusion

If you understood the major concepts behind all the chapters in this book, you can call yourself the 'Macro expert'. But we would encourage you to practice the concepts conveyed in this book. By practicing, you will better grasp complex concepts. As they say, 'Practice makes a man perfect'. If you want to make yourself an expert and perfect candidate for 'Excel Macro' jobs then you have to practice the concepts repeatedly to get a strong grip on them.

In this book, we tried our best to explain concepts at the very basic level. The purpose of explaining each and every step at the basic level was to make a layman understand what Macros are and what are their applications and how to apply them in real life. We tried our best to cover all the major topics regarding 'Excel Macros' in this book.

EXCEL

FORMULAS

AND FUNCTIONS

For Complete Beginners, Step-By-Step Illustrated Guide to Master Formulas and Functions

Introduction

Spreadsheets have been with us for a long time. The best-known and most widely used spreadsheet is Microsoft Excel. Excel is easy to use for most daily number crunching tasks and it comes with formulas and functions that perform a host of tasks. Tasks such as summing, manipulating text, averaging, comparing, answering what if questions...

Assuming you know the meaning of such basic spreadsheet terms, as row, column, and cell, and you know how to do such basic operations as copy and paste, this little book will introduce you to a number of the formulas and functions of Excel. It will give you the ability and confidence to use them successfully.

Chapter 1: What are formulas and functions?

A **formula** is an expression used to calculate the value in a cell.

For example, here is a very simple problem where A2 contains 5.7 and B2 contains 6.4. The task is to put the products of these two numbers in into cell C2. See diagram below.

A	B	C
5.7	6.4	

We can do this by selecting c2 and writing the very simple formula = **A2*B2** then pressing enter [it does not matter if you write **a2*B2, A2*b2** or **a2*b2**, as uppercase and lower case letters are treated the same when referring to columns].

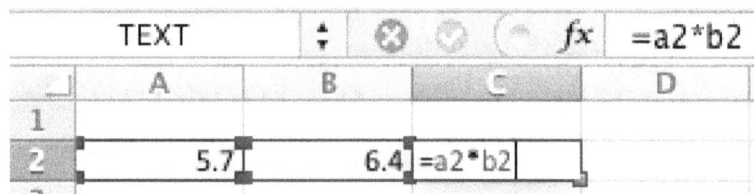

When this is done it leads to the value of the product appearing in the cell C2. [The word 'TEXT' and the expression *fx* = A2*B2, which are shown in the diagram appear automatically. You can ignore them.]

| 2 | 5.7 | 6.4 | 36.48 |

Another simple example is adding the contents of the cells A2, A3, A4, and putting the results in A5.

	A
1	
2	5.7
3	7.8
4	10.3
5	

Again, we select A5 and write = **A2+A3+A4**.

	A
1	
2	5.7
3	7.8
4	10.3
5	=a2+a3+a4

After we press enter the result is obtained and ends up in A5.

	A
1	
2	5.7
3	7.8
4	10.3
5	23.8

So far so good, but what if you needed to add the contents of 100 cells. It would take a long time to write = **A2+A3+A4+ A5...... +A100 + A101** and you would probably make an error.

97	51
98	28
99	94
100	52
101	34

Fortunately, the creators of Excel anticipated many needs of users decades ago and created functions. For this particular problem, we use the sum function.

Select cell c102 and type, '=
SUM(a2:a101) '[this is called the function
declaration].

97	51
98	28
99	94
100	52
101	34
102	=SUM(a2:a101)
103	

then press enter and cell a102 has the sum,
which in this case was 6086.

97	51
98	28
99	94
100	52
101	34
102	6086

A **function** is a predefined formula, which
performs calculations with the contents of
cells.

Parameters

You may notice in the expression,
'=SUM(a2:a101**)'** the interesting phrase
a2:a101. This is the *range* of the function.
The range of a function is the set of cells on
which it acts. The range of the SUM
function is called its *parameter*, as it can
vary. We will have different ranges to which
the SUM function is applied.

There are a vast number of functions in
Excel, which you will become familiar with
as you master the use of this amazing tool.
This little book will look at some of the
more important functions.

Chapter 2: Text Formulas

The Excel **TEXT** formula or function is a most interesting one.

Date format

 The Excel TEXT formula or function can be used to bring about changes of date and number format. Here is an example, suppose we wanted to change a column of dates in a certain format

2	19-Mar-01
3	20-Mar-01
4	21-Mar-01
5	22-Mar-01
6	23-Mar-01
7	24-Mar-01
8	25-Mar-01

and wish to change them into the form mm/dd/yyyy. In order to do this, we use the Text function.

We can pick any column of 7 cells to put them in, so let's put them in the column starting at a2. Select a2 and type the function declaration,'

=TEXT(b2,"mm/dd/yyyy") '

	TEXT		⊗	⊘		fx	=TEXT(b2,"mm/dd/yyyy")		
	A	B		C		D	E		F
1									
2	=TEXT(b2,"mm/dd/yyyy")								
3	TEXT(value, format_text)								
4		21-Mar-01							
5		22-Mar-01							
6		23-Mar-01							
7		24-Mar-01							
8		25-Mar-01							

If you do this, Excel automatically offers you possible functions to make your task easier. You are sensible to use them.

Once you filled the function in press enter and you will find the correct expression in a2.

You can quickly fill in the rest of the column by using the 'handle' at the side of a2.

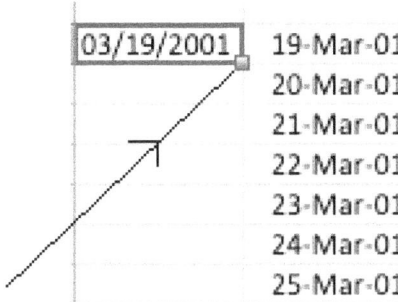

	19-Mar-01
03/19/2001	20-Mar-01
	21-Mar-01
	22-Mar-01
	23-Mar-01
	24-Mar-01
	25-Mar-01

Then drag down.

2	03/19/2001	19-Mar-01
3	03/20/2001	20-Mar-01
4	03/21/2001	21-Mar-01
5	03/22/2001	22-Mar-01
6	03/23/2001	23-Mar-01
7	03/24/2001	24-Mar-01
8	03/25/2001	25-Mar-01

Currency

Another possible use of the **TEXT** function is to convert a column of numbers

E
5
6
78.9
23.97

into currency format.

Note the first number is contained in e2. Suppose the new column will start at f10. Select f10 and type the function declaration, ' = **TEXT(e2,"$##.##")** '

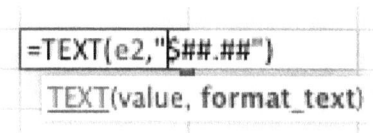

Press enter.

Now use the handle and move down to get the remaining conversions.

$5.
$6.
$78.9
$23.97

TEXT can be used in many ways like the two examples shown.

If we look at the structure or syntax of the TEXT function through a declaration, **'=TEXT(e2,"$##.##")'**, we see the parameters of the TEXT function are TEXT(value, format_text). Value, as the contents of cells, is values; format_text, as the function acts to change the format of the value.

Chapter 3: Comparison Formulas

Sometimes you wish to compare two columns.

87	64
57	28
92	23
81	33
64	14
56	91

You test whether two specific cells match with the IF function [We will say more about this later]. Begin a formula with =**IF(** and enter the two cell locations, with = between them. Put in a comma then enter relevant text, in quotes, to show if the cells match. Put in another comma and then the relevant text, in quotes, if there is a non-match.

87	64	=IF(I11=J11,"Match","No Match")
57	28	IF(logical_test, [value_if_true], [value_if_false])
92	23	
81	33	
64	14	
56	91	

Now press **Enter** to complete the formula.

87	64	**No Match**
57	28	
92	23	
81	33	
64	14	
56	91	

Use the handle to complete the comparison.

87	64	**No Match**
57	28	**No Match**
92	23	**No Match**
81	33	**No Match**
64	14	**No Match**
56	91	**No Match**

Chapter 4: Operators

Excel has operators in four categories. These are arithmetic operators, comparison operators, text concatenation operators, and reference operators.

The arithmetic operators are the usual +, -, × and ÷ operators except that in Excel × is written as * and ÷ as/. There are two other important arithmetic operators. These are % [percentage, which calculates a percentage] and ^ [caret, this raises to powers. See below].

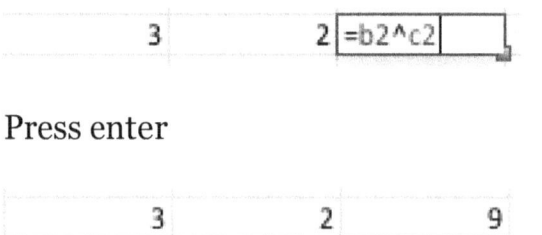

Press enter

3	2	9

The comparison operators are as they are in arithmetic and they are: >, < , ≤ , ≥ , =, ≠, however Excel uses <= for ≤ , >= for ≥ and <> for ≠.

Text concatenation may be a new idea for some of you. There is only one text concatenation operator and that is & [ampersand]. Here is how it works, suppose you had the string [word], "nice" in a2 and "cat" in b2

	A	B	C
1			
2	nice	cat	

then = **a2 & b2** gives "nicecat".

	A	B	C
1			
2	nice	cat	= a2&b2

	A	B	C
1			
2	nice	cat	nicecat

135

All this operator does is join words or strings into one word or string.

 Another type of operator, which may be new to you is the type referred to as a reference operator. Here are the reference operators.

: is the colon operator and refers to all references in a range from one cell to another including the endpoints.

 Example. a2:a4 refers to a2, a3, and a4.

, is the comma operator and joins two ranges together. Here is an example a2: a4, b2: b4 refers to a2, a3, a4, b2, b3, b4. It is the equivalent in mathematics of the union operator [∪].

Finally, we have the single space operator. This is just an empty space. It produces references to common cells and ranges.

Example. a2:a4 a3:a5 gives a3, a4. Obviously, a3 and a4 are the only common cells of the ranges a2, a3, a4 and a3, a4, a5. It is the equivalent in mathematics of the intersection operator [∩].

A very important consideration in constructing formulas is the precedence or order of operators, in order to get the correct result. All formulas begin with = and if you are dealing with numbers the order is similar to the order of arithmetical operators as described by BEDMAS [Brackets →Exponents → Division and Multiplication →Addition and Subtraction] or PEMDAS [Brackets →Exponents →Multiplication

and Division →Addition and Subtraction]
taught in high school.

Here is how the precedence of Excel
operations works.

: [colon], single-space, , [comma], negation
[as in -7], % [percentage], ^ [caret], *
and/[neither is above the other], + and -
/[neither is above the other], &
[ampersand], =, <,>,<=, >=,<>
[comparison].

Chapter 5: Absolute vs Relative Cell References

When you working with Excel, you must know about what is called relative vs. absolute cell reference.

The problem is this: if you COPY A FORMULA containing cell references, generally the CELL REFERENCES CHANGE!

The diagrams below illustrate this.

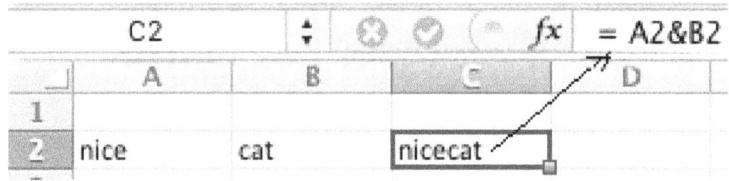

Now copy these three cells to a1,b1,c1 and look what happens.

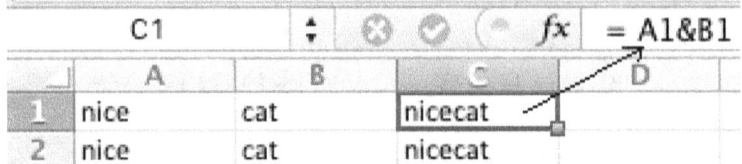

	A	B	C	D
1	nice	cat	nicecat	
2	nice	cat	nicecat	

C1   *fx* = A1&B1

These cell references are named "relative" cell references, as they change relative to where the formula is copied.

Sometimes, you don't want cell references altering when a formula is copied. If that is the case then you use what is called absolute cell references. In order to use absolute cell references, you put a "$" before the column letter if you want that to always remain the same. Similarly, you put a "$" in front of the row number if you want that to remain the same.

The following diagrams should help you grasp this very important idea.

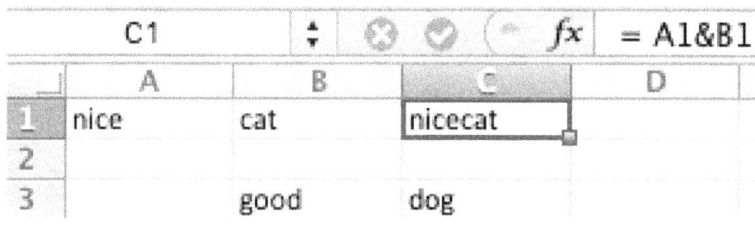

In this diagram, we are using relative references. Now copy the cell C2 to D3.

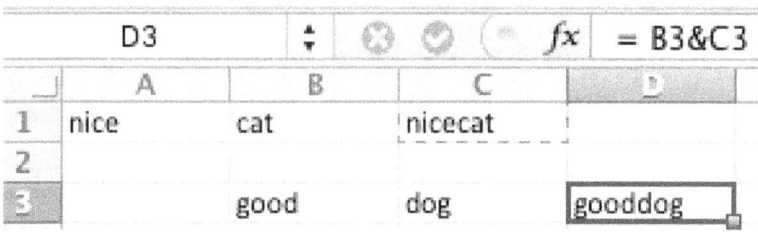

Note how the cells have changed from A1→B3 and B1→C3.

Now let's do this using Absolute References.

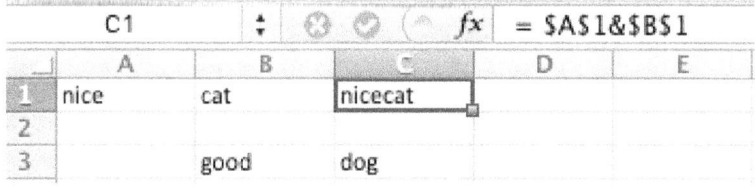

The first diagram is as we got last time because "nice" is actually in the cell A1 and "cat" is actually in B1. Now let's copy the cell C2 to D3, as we did before

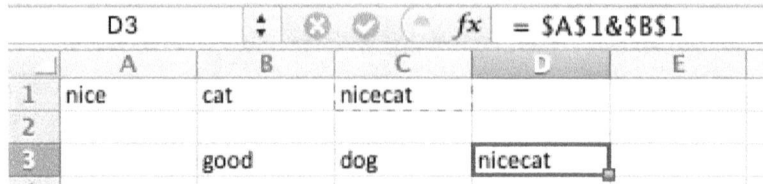

	A	B	C	D	E
				fx = $A$1&$B$1	
1	nice	cat	nicecat		
2					
3		good	dog	nicecat	

This time we got 'nicecat' instead of 'gooddog' because the reference was to A1 and B1 exactly and not just to the two cells immediately to the left of the cell we were copying.

When you create spreadsheets decide before copying a formula the cell references that are relative and those that are absolute.

You will realize the absolute importance of these ideas when we look at the VLOOKUP and HLOOKUP functions later.

Chapter 6: SUM

We had a look at the SUM function in chapter 1. If you have forgotten this have another read. However, here is another look at this extraordinary function.

This time, we're going to sum 3 columns of numbers.

A	B	C	D
	66	57	42
	64	77	49
	28	71	85
	23	59	61
	33	47	28
	14	50	88
	91	57	41
	87	53	88
	57	78	71
	92	81	16

In order to do this, we shall use the : and , operators discussed in the last chapter. In D12, write,' **=SUM(b2:b11,c2:c11,**

d2:d11) ' then press enter. The sum of
these numbers is 1754.

	D12		fx	=SUM(B2:B11,C2:C11,D2:D11)		
	A	B	C	D	E	F
1						
2		66	57	42		
3		64	77	49		
4		28	71	85		
5		23	59	61		
6		33	47	28		
7		14	50	88		
8		91	57	41		
9		87	53	88		
10		57	78	71		
11		92	81	16		
12				1754		

Chapter 7: IF

In many ways, the IF function is the most important function Excel has. Earlier, we showed how it could be used for comparisons. Here is another example of this. We are going to have two columns of numbers. Using IF, a column to the right will have the word "CAT" if the number in the left column is no less than the number in the right. If this condition is not met, in other words, if the number in the left column is less than the number in the right then we will get "DOG". Here are the columns.

	A	B	C
1			
2		88	57
3		42	51
4		57	58
5		77	42
6		71	49
7		59	85
8		47	61
9		50	28
10		57	88
11		53	41
12		78	88
13		81	71
14		81	16
15		61	56
16		53	59

Into cell a2 type the function declaration,' =
IF(b2>=c2, "CAT","DOG") ' then press
enter.

D2				fx	=IF(B2>=C2,"CAT","DOG")	
	A	B	C	D	E	F
1						
2		88	57	CAT		
3		42	51			

Now, you can use the 'handle' to complete
the use of the function.

146

	D4				fx	=IF(B4>=C4,"CAT","DOG")	
	A	B	C	D		E	F
1							
2		88	57	CAT			
3		42	51	DOG			
4		57	58	DOG			
5		77	42	CAT			

Now before leaving the IF function, we will consider another extremely useful property of this extraordinary function:- NESTING. In this context, nesting has nothing to do with birds. It is the ability of the IF function to make itself one of the choices. Here is an example to show this.

Example. We are given the following column of numbers.

	A	B	C
1			
2		188	
3		1142	
4		57	
5		77	
6		1	
7		459	
8		47	
9		504	
0		257	
1		6753	
2		798	
3		81	
4		281	
5		61	
6		5	

Into C column beside the numbers in B column, there will be sentences saying," Bi has a number with x digits." Thus by 188, there will be, "188 has 3 digits. The next diagram will show how this is done using the IF function.

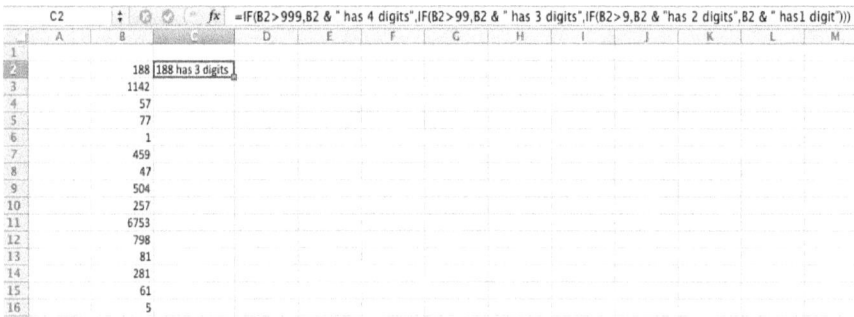

Using the 'handles' solves the final part of the problem.

188	188 has 3 digits
1142	1142 has 4 digits
57	57has 2 digits
77	77has 2 digits
1	1 has1 digit
459	459 has 3 digits
47	47has 2 digits
504	504 has 3 digits
257	257 has 3 digits
6753	6753 has 4 digits
798	798 has 3 digits
81	81has 2 digits
281	281 has 3 digits
61	61has 2 digits
5	5 has1 digit

Structure or Syntax of IF

If we look at the structure or syntax of the IF function through a declaration, ' = **IF(b2>=c2, "CAT","DOG")** ', we see the parameters of the IF function are IF(logical test, result if TRUE, result if FALSE). Logical test, which is either TRUE or FALSE, as the contents of cells are values; Result if TRUE; Result if FALSE.

Chapter 8: AND

Another widely used Excel function is the AND function. It is either TRUE or FALSE depending on whether two conditions are both true or not.

Here is an example, which illustrates this. We need to compare the sizes of the numbers contained in the rows of this table using the AND function.

	A	B
1		
2	23	90
3	56	78
4	34	48
5	55	12

Into cell C2 type the function declaration,'
=AND(A2<50,B2<50)'

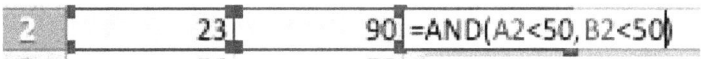

then press enter.

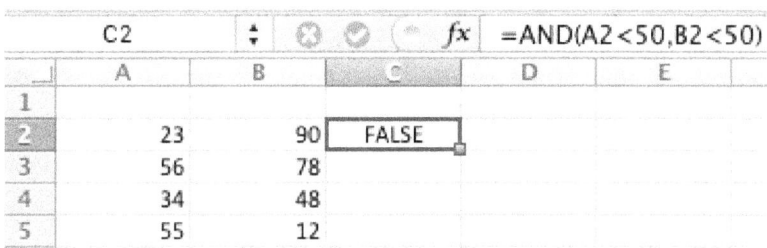

Now use the handle to complete the task

2	23	90	FALSE
3	56	78	FALSE
4	34	48	TRUE
5	55	12	FALSE

The only row, which had A and B cells both
less than 50 was row 4.

Here is another example of the same type. Find the rows in which the C and D cells are both less than 50 in the table below.

C	D
14	57
12	22
87	71
57	9
92	47
16	50
64	57
56	53
17	22
46	81
88	23

Into cell E2 type the function declaration, ' **=AND(C2<50,D2<50)** ' then press enter.

	fx	=AND(C2<50,D2<50)	

C	D	E
14	57	FALSE
12	22	
87	71	
57	9	
92	47	
16	50	
64	57	
56	53	
17	22	
46	81	
88	23	

Now use the handle to complete the task.

14	57	FALSE
12	22	TRUE
87	71	FALSE
57	9	FALSE
92	47	FALSE
16	50	FALSE
64	57	FALSE
56	53	FALSE
17	22	TRUE
46	81	FALSE
88	23	FALSE

In this case, only rows 3 and 10 had C and D cells where the number was less than 50.

Structure or Syntax of AND
If we look at the structure or syntax of the AND function through a declaration, '**=AND(C2<50, D2<50)**', we see the parameters of the AND function are AND(logical test, logical test). The parameters are two logical tests, the combination of whose results give different results.

Chapter 9: LEN

The LEN function is one of a large number of Excel functions, which handle strings [words]. The LEN function has a very simple format = **LEN (**string**)**. The function returns the number of letters in a string, which is often in a cell.

Here is an example. We have a column of names and need to write beside it the number of letters in each of the names.

The input in cell B2 is = LEN [A2].

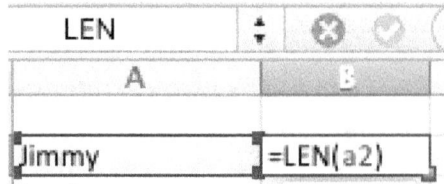

Now press enter. This action produces

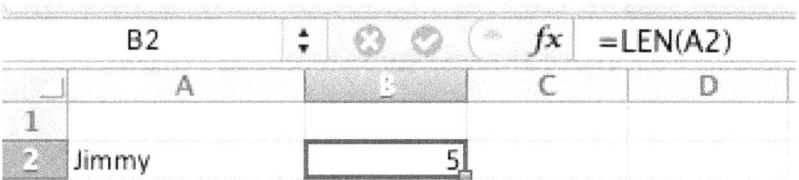

B2		⊗ ⊘	fx	=LEN(A2)
	A	B	C	D
1				
2	Jimmy	5		

Then we use the handle to produce

D21	⊗ ⊘
A	B
Jimmy	5
Jamie	5
Marsden	7
Oliver	6
Sean	4
Jaxon	5
Bart	4
Daniel	6
Dominic	7
Alex	4
Leo	3
Vinu	4
Harry	5
Vincent	7
Oliver	6
Nathan	6
Jean-Michael	12
Scott	5
Matthew	7
Siddhartha	10

There are many string functions in Excel. With Excel string functions, you can do all sorts of things, such as joining text from different cells to make a new one string, removing parts of a string depending on their position or surroundings, substitution into parts of a string, etc. Sometimes, these functions are referred to as text functions. This sometimes leads to confusion with the TEXT function, which we looked at in Chapter 2 .

The LEN function is a string function, some other very important string functions are the LEFT and Right Functions. The LEFT function gets a substring that contains a given *number of left characters* from a string. Similarly, the RIGHT function gets a substring that contains a given *number of right characters* from a string.

The following example shows how the LEFT function works.

Example. There is a column of names from which we want the two most left characters.

The input in cell B2 is = LEFT (A2,2).

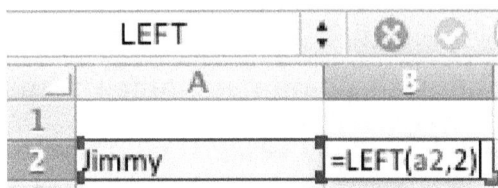

Now press enter. This action produces

| Jimmy | Ji |

Then we use the handle to produce

	A	B
1		
2	Jimmy	Ji
3	Jamie	Ja
4	Marsden	Ma
5	Oliver	Ol
6	Sean	Se
7	Jaxon	Ja
8	Bart	Ba
9	Daniel	Da
10	Dominic	Do
11	Alex	Al
12	Leo	Le
13	Vinu	Vi
14	Harry	Ha
15	Vincent	Vi
16	Oliver	Ol
17	Nathan	Na
18	Jean-Michael	Je
19	Scott	Sc
20	Matthew	Ma
21	Siddhartha	Si

The RIGHT function is used in an identical way.

The Upper Function is a simple function UPPER(string), which acts to convert all letters or characters in a string to upper case.

Here is an example of its use, UPPER('cat') = CAT.

Here is the result of its use combined with the handle on a column of names.

There are many more string functions in Excel but we will only look at one more, which is the MID function.

The structure or syntax of the function is MID(*string*, *number1*, *number2*).

What do these parameters mean?

The *string* is the string or word, which the function is acting on, *number1* is the number of the letter in the string that you start with and *number2* is the number of the letters in the string that you want, starting with the letter at *number1*.

Let's take the word, 'ELEPHANT' and calculate MID('ELEPHANT', 2, 5). The letter at position 2 is L and the fifth letter after starting with L is A, hence MID('ELEPHANT', 2, 5) = LEPHA.

Now let's use the MID function on a column of strings.

B1		⊗ ⊘	fx	=MID(A1, 1,2)
	A	B	C	D
1	Pan	Pa		
2	Suen	Su		
3	Chandrakumar	Ch		
4	Lee	Le		
5	Wu	Wu		
6	Aitchison	Ai		
7	Bell	Be		
8	Hawthorne	Ha		
9	Johnson	Jo		
10	Phillips	Ph		
11	Vale	Va		
12	Webber	We		
13	Brown-Bayliss	Br		
14	Daffron	Da		
15	Beynon	Be		

Chapter 10: OR

A widely used Excel function similar to the AND function is the OR function. It is either TRUE or FALSE depending on whether one or both of two conditions are true or not.

Here is an example, which illustrates this.

Example. Find the rows in which **either** the A and B cells are less than 50 in the table below.

	A	B
1		
2	53	49
3	78	64
4	23	28
5	81	23
6	61	33
7	53	14
8	57	91
9	51	87
10	58	57
11	42	92

E13

Into cell C2 type the function declaration,'
=OR(A2<50,B2<50)' then

| 2 | 53 | 49 =OR(a2<50,b2<50) |

then press enter.

E14		✕ ✓ fx
A	B	C
53	49	TRUE
78	64	

Finally, complete the task by using the handle.

	A	B	C
1			
2	53	49	TRUE
3	78	64	FALSE
4	23	28	TRUE
5	81	23	TRUE
6	61	33	TRUE
7	53	14	TRUE
8	57	91	FALSE
9	51	87	FALSE
10	58	57	FALSE
11	42	92	TRUE

In this case, we get TRUE for rows 2, 4, 5, 6, 7 and 11, as either one or both the A and B cells contain numbers, which are less than 50.

Structure or Syntax of OR
If we look at the structure or syntax of the OR function through a declaration, '
=OR(A2<50, B2<50)', we see the parameters of the OR function are OR(logical test, logical test). The parameters

are two logical tests, the combination of whose results give different results.

Chapter 11: NOT

The NOT function is an interesting one. It only has a meaningful result if a cell has a data type called Booleans, which are either TRUE or FALSE. The number 1 is regarded as TRUE and 0 as FALSE.

The function NOT gives a value of FALSE if the cell contains TRUE and TRUE if the cell contains FALSE.

Here is an example. In the last chapter, we got the table.

D19

	A	B	C
1			
2	53	49	TRUE
3	78	64	FALSE
4	23	28	TRUE
5	81	23	TRUE
6	61	33	TRUE
7	53	14	TRUE
8	57	91	FALSE
9	51	87	FALSE
10	58	57	FALSE
11	42	92	TRUE

Use the NOT function on column C.

Into cell D2 type the function declaration

=NOT(C2)

| 2 | | 53 | | 49 | TRUE | =NOT(c2) |

then press enter.

D12				fx	
A	**B**	**C**		**D**	
53	49	TRUE		FALSE	
78	64	FALSE			
23	28	TRUE			
81	23	TRUE			

Now finish the task by using the handle.

	A	B	C	D
1				
2	53	49	TRUE	FALSE
3	78	64	FALSE	TRUE
4	23	28	TRUE	FALSE
5	81	23	TRUE	FALSE
6	61	33	TRUE	FALSE
7	53	14	TRUE	FALSE
8	57	91	FALSE	TRUE
9	51	87	FALSE	TRUE
10	58	57	FALSE	TRUE
11	42	92	TRUE	FALSE

You may well ask what effect the NOT function has on cells, which contain data apart from Boolean. The table below gives the results on some such cells.

	B9	⬍	❌ ✅	fx	=NOT(A9)

	A	B	C	D
1				
2	1	FALSE		
3	0	TRUE		
4	3	FALSE		
5	cat	#VALUE!		
6	FALSE	TRUE		
7	45	FALSE		
8	23	FALSE		
9	117	FALSE		

It would seem any number except 0 is considered TRUE. Text like 'cat' is not recognized.

Only use the NOT function on Boolean data!

Chapter 12: XOR

A widely used Excel function similar to the OR function is the XOR function. Like the OR function, it is either TRUE or FALSE depending on whether one or both of two conditions are true or not. It is fussier than OR, as it is TRUE only if one of the conditions is TRUE. If both conditions are TRUE then the XOR function gives False.

Here is an Example. We want TRUE if **only one** of Family or First name has a length less than 7.

H	I
Family	First
Benton	Jimmy
Brooke	James
Cheong	Marshan
Griffith	Sam
Hicke	Sean
Howan	June
Jorgensen	Bob
McAndrew	Diana
Mildeng	Dominic
Pang	Alex

Into cell J2 type the function declaration,'
=XOR(LEN(H2)<7, LEN(I2)<7) ' press
enter then use the handle to get

Family	First	
Benton	Jimmy	FALSE
Brooke	James	FALSE
Cheong	Marshan	TRUE
Griffith	Sam	TRUE
Hicke	Sean	FALSE
Howan	June	FALSE
Jorgensen	Bob	TRUE
McAndrew	Diana	TRUE
Mildeng	Dominic	FALSE
Pang	Alex	FALSE

The reason that 'Benton Jimmy' gives
FALSE is that the lengths of both 'Benton'
and 'Jimmy' are less than 7 and with XOR
this will give FALSE, whereas if we had
used OR we would have got TRUE.

Structure or Syntax of XOR
We could look at the structure or syntax of the XOR function through a declaration, '**=XOR(A2<50, B2<50)**' in the same way that we did for the AND and OR functions. If we do so, we see the parameters of the XOR function are XOR(logical test, logical test). The parameters are two logical tests; the combination of whose results give different results.

NB
The XOR function is a new one that was not in versions of Excel prior to 2013.

Chapter 13: SUMIF and SUMIFs

These functions are very similar to the SUM function except that they include the ability to put criteria in the function declaration. The best way to show this is with some examples.

Here is the first example. In the table of data below, we want to add all grades less than 50, using the functions of Excel.

H3			fx	
A	B	C	D	
Name	First	Gender	Grade	
Benson	Jane	F	66	
Brook	James	M	64	
Cheong	Bob	M	28	
Griffin	Otis	M	23	
Hicks	Sally	F	33	
Howan	Jill	F	14	
Jorgensen	Bob	M	91	
McAndrew	Diana	F	87	
Mildon	David	M	57	
Pang	Alex	M	92	
Suen	Luke	M	81	
Chandrakum	Sakib	M	64	
Lee	Helen	F	56	
Wu	Valeri	F	78	

Quite clearly the sum required is 28 + 23 + 33 + 14 = 98. Let's get this using Excel.

Into any cell, except those with the data, type the function declaration, '=SUMIF(D2: D15,"<50", D2: D15) ' then press enter and you get 98.

Interestingly, as the range with the criterion < 50 is the same as the range we're summing over, we don't need the second D2: D13.

We would get exactly the same result if the function declaration was, '=SUMIF(D2:D15,"<50") '. Try it and see for yourself.

If the range with the criterion is different from the range we're summing over this is not true.

Example. Find the sum of all grades of Female students.

	A	B	C	D
H3		fx		
	Name	First	Gender	Grade
	Benson	Jane	F	66
	Brook	James	M	64
	Cheong	Bob	M	28
	Griffin	Otis	M	23
	Hicks	Sally	F	33
	Howan	Jill	F	14
	Jorgensen	Bob	M	91
	McAndrew	Diana	F	87
	Mildon	David	M	57
	Pang	Alex	M	92
	Suen	Luke	M	81
	Chandrakum	Sakib	M	64
	Lee	Helen	F	56
	Wu	Valeri	F	78

Into any cell, except those with the data, type the function declaration, ' =SUMIF(C2:C14,"F",D2:D15) ' then press enter. This time the result is 334.

The only trouble with SUMIF is that it only allows one criterion. If you want multiple criteria [plural of criterion] then you have to use SUMIFS. Once again an example will show what this entails.

Example. Find the sum of the grades of males whose school is 'A'.

	D25			fx	
	A	B	C	D	E
1	Name	First	Gender	School	Grade
2	Benson	Jane	F	A	66
3	Brook	James	M	A	64
4	Cheong	Bob	M	B	28
5	Griffin	Otis	M	A	23
6	Hicks	Sally	F	B	33
7	Howan	Jill	F	A	14
8	Jorgensen	Bob	M	A	91
9	McAndrew	Diana	F	B	87
10	Mildon	David	M	B	57
11	Pang	Alex	M	B	92
12	Suen	Luke	M	A	81
13	Chandrakum	Sakib	M	B	64
14	Lee	Helen	F	B	56
15	Wu	Valeri	F	B	78

Into any cell, except those with the data, type the function declaration, '
=SUMIFS(E2:E15,C2:C15,"M", D2:D15,"A")

' then press enter. This time the result is 259.

The declaration is,' =SUMIFS(sum range, criterion 1, criterion 2, criterion 3,....)'. Sum range is always like E2: E15 and is the set you want to add over, criteria are always like D2: D15, "A", where you have a range where the criterion applies then a comma then the actual criterion.

These two functions have a structure or syntax so that the parameters are range and logical tests.

Chapter 14: COUNT and COUNTA

This chapter deals with two very simple functions: COUNT and COUNTA. All the COUNT function does is count the cells in a range, which contain numbers. The COUNTA function, however, counts all non-empty cells.

Example. Count all the cells with numbers only in the cells below.

M	N
Benson	McAndrew
56	Mildon
Cheong	Pang
45	Suen
78	33
Howan	90
Jorgensen	Wu
	7

Into any cell, except those with the data, type the function declaration, '

=COUNT(M2: M9, N2: N9) ' then press enter. The result is 6, as there six cells, which contain numbers.

Now let's use COUNTA. Into any cell, except those with the data, type the function declaration, ' =COUNTA(M2: M9, N2: N9) ' then press enter. The result is 15, as there fifteen cells, which contain numbers. Note that the empty cell is not counted, either by COUNT or COUNTA.

These two functions have a very simple structure or syntax. Their only parameter is a range.

Chapter 15: AVERAGEIF and AVERAGEIFs

The AVERAGEIF and AVERAGEIFS functions are very similar to the SUMIF and SUMIFS functions.

Example. Suppose you wanted to find the average grade of the female students in the table below.

	A	B	C	D	E
	K31			fx	
1	Name	First	Gender	School	Grade
2	Benson	Jane	F	A	66
3	Brook	James	M	A	64
4	Cheong	Bob	M	B	28
5	Griffin	Otis	M	A	23
6	Hicks	Sally	F	B	33
7	Howan	Jill	F	A	14
8	Jorgensen	Bob	M	A	91
9	McAndrew	Diana	F	B	87
10	Mildon	David	M	B	57
11	Pang	Alex	M	B	92
12	Suen	Luke	M	A	81
13	Chandrakum	Sakib	M	B	64
14	Lee	Helen	F	B	56
15	Wu	Valeri	F	B	78

Into any cell, except those with the data, type the function declaration, '
=AVERAGEIF(C2: C15, "F", E2: E15) ' then press enter. This time the result is 55.666....

Generally, the declaration is, '
=AVERAGEIF(criterion range, criterion, average_range) '. The meaning of average_range is the range you're going to take the average of. AVERAGEIF is for situations where there is only one criterion.

If you have multiple criteria then you must use AVERAGEIFS.

Example. Suppose you wanted to find the average grade of the male students who went to school B from the table below.

	A	B	C	D	E
1	Name	First	Gender	School	Grade
2	Benson	Jane	F	A	66
3	Brook	James	M	A	64
4	Cheong	Bob	M	B	28
5	Griffin	Otis	M	A	23
6	Hicks	Sally	F	B	33
7	Howan	Jill	F	A	14
8	Jorgensen	Bob	M	A	91
9	McAndrew	Diana	F	B	87
10	Mildon	David	M	B	57
11	Pang	Alex	M	B	92
12	Suen	Luke	M	A	81
13	Chandrakum	Sakib	M	B	64
14	Lee	Helen	F	B	56
15	Wu	Valeri	F	B	78

Into any cell, except those with the data, type the function declaration, ' =AVERAGEIFS(E2:E15,C2:C15,"M", D2:D15,"B") ' then press enter. This time the result is 60.25

Generally, the declaration is, ' =AVERAGEIFs(average_range, criterion1 range, criterion1, criterion2 range, criterion2,) '. The meaning of average_range is the range you're going to take the average of. AVERAGEIFS is for situations where there is more than one criterion.

Chapter 16: LARGE and SMALL

The LARGE and SMALL functions are very easy to understand and use. The declaration ' = LARGE(F1: G100, 3)' finds the third largest number in the range F1: G100.

Example. Find the fifth largest number of the set in the cells below.

J	K
66	78
64	56
28	97
23	112
33	0.8
14	79
91	67
87	
57	
92	
81	
64	
56	
78	

Into any cell, except those with the data, type the function declaration, '

=LARGE(J2: K8, 5) ' then press enter. The result is 78.

Example. Find the second smallest number of the set in the cells below.

J
66
64
28
23
33
14
91
87
57
92
81
64
56
78

Into any cell, except those with the data, type the function declaration, '
=SMALL(J2: J15, 2) ' then press enter. The result is 23.

Chapter 17: COUNTIF and COUNTIFS

The COUNTIF and COUNTIFS functions are very similar to the SUMIF and SUMIFS functions.

Example. Suppose you wanted to find female students were in the table below.

K31					fx	
	A	B	C	D	E
1	Name	First	Gender	School	Grade
2	Benson	Jane	F	A	66
3	Brook	James	M	A	64
4	Cheong	Bob	M	B	28
5	Griffin	Otis	M	A	23
6	Hicks	Sally	F	B	33
7	Howan	Jill	F	A	14
8	Jorgensen	Bob	M	A	91
9	McAndrew	Diana	F	B	87
10	Mildon	David	M	B	57
11	Pang	Alex	M	B	92
12	Suen	Luke	M	A	81
13	Chandrakum	Sakib	M	B	64
14	Lee	Helen	F	B	56
15	Wu	Valeri	F	B	78
1 C					

Into any cell, except those with the data, type the function declaration, '
=COUNTIF(C2: C15, "F") ' then press enter. The result is 6

Generally, the declaration is, '
=COUNTIF(criterion range, criterion) '. COUNTIF is for situations where there is only one criterion.

If you have multiple criteria then you must use COUNTIFS.

Example. Suppose you wanted to find the number of male students who went to school B from the table below.

	A	B	C	D	E
	K31			fx	
1	Name	First	Gender	School	Grade
2	Benson	Jane	F	A	66
3	Brook	James	M	A	64
4	Cheong	Bob	M	B	28
5	Griffin	Otis	M	A	23
6	Hicks	Sally	F	B	33
7	Howan	Jill	F	A	14
8	Jorgensen	Bob	M	A	91
9	McAndrew	Diana	F	B	87
10	Mildon	David	M	B	57
11	Pang	Alex	M	B	92
12	Suen	Luke	M	A	81
13	Chandrakum	Sakib	M	B	64
14	Lee	Helen	F	B	56
15	Wu	Valeri	F	B	78

Into any cell, except those with the data, type the function declaration, '
=COUNTIFS(C2:C15,"M", D2:D15,"B") '
then press enter. This time the result is 4.

Generally, the declaration is, '
=COUNTIFS(criterion1 range, criterion1,

191

criterion2 range, criterion2,....) '.
COUNTIFS is for situations where there is more than one criterion.

Chapter 18: VLOOKUP

The next two chapters deal with the VLOOKUP and HLOOKUP functions. These are very useful but are often misunderstood.

We will need a number of examples to try and make this clear. Here is the first example.

We have a table of grades from a school.

	A	B	C	D	E
	Family	First	Gender	English	Math
1	Family	First	Gender	English	Math
2	Argentin	Brady	M	56	71
3	Brewer	Shane	M	34	49
4	Jippu	Steven	M	80	88
5	Lattimore	Zelda	F	76	85
6	Liu	Phillippa	F	45	28
7	Martelil	Matthew	M	48	50
8	McTigure	Panama	F	65	57
9	Nichollson	Carla	F	95	88
10	Olsen	Brooklyn	F	48	53
11	Prapaithong	Jill	F	65	61
12	Ramez	Dillon	M	72	78
13	Sharo	William	M	35	41
14	Singh	Sakib	M	78	81
15	Stead	Helen	F	45	81
16	Stuart	Janine	F	56	61
17	Thompson	Mary	F	34	53
18	Wang	Maxine	F	80	57
19	Watson	Chloe	F	33	51
20	Wong	Vince	M	56	58
21	Zhang	Sue	F	67	42

The formula bar shows: E36 and fx

and we want to put the math grade beside each Family Name in the list below.

H	I
Martelil	
Stuart	
Ramez	
Olsen	
McTigure	
Stead	
Singh	

This could be done by the extremely tedious procedure of copying from the table of grades to this table. If we did this Martelil would have 50 beside him and so on. Fortunately, the creators of Excel foresaw this problem and devised the LOOKUP functions.

Here is how this problem is solved using the VLOOKUP function.

Into I2 type, ' =

VLOOKUP(H2,$A$2:$E$21, 5, FALSE) '

then press enter. The result is the value 50.

H	I
Martelil	50
Stuart	
Ramez	
Olsen	
McTigure	
Stead	
Singh	

We finish the procedure with the handles.

H	I
Martelil	50
Stuart	61
Ramez	78
Olsen	53
McTigure	57
Stead	81
Singh	81

Obviously, VLOOKUP works, but what does each part of VLOOKUP(H2,$A$2:$E$21, 5, FALSE) mean?

- H2 is the cell whose corresponding math grade is going to be looked up.
- $A$2:$E$21 is the range of values from which we will look up the math grade. Note the use of absolute references. This is usually **ESSENTIAL!**
- 5 refers to the column in the range where we get math grades. Column 1 is taken as the LOOKUP column. Its data must be in ASCENDING order.
- FALSE means we want an exact match between the names in our list and in the column Family. If we had used TRUE then the function would have been satisfied with an approximate match.

Finally, and these are extremely important:

1. **The column we are looking for matches in must be the first column in the range.**
2. **The first column must be put in ascending order**.

If you don't know how to order data then make sure to read the last chapter.

Now another example, which will hopefully help cement these ideas.

A company has 5 employees who are paid different hourly rates, as shown below in a Sheet1 of a spreadsheet using the range A1: B6.

	A13		

	A	B
1	Name	Hourly Rate
2	Alan	$40
3	Bob	$45
4	Cathy	$50
5	Diane	$35
6	Ed	$35

In order to pay these people, it is necessary to fill in the Hourly Rate in the table below in Sheet2 of the same spreadsheet.

	C15			fx

	A	B	C
1	Name	Hours	Hourly Rate
2	Diane	42	
3	Ed	36	
4	Alan	43	
5	Bob	29	
6	Cathy	56	
7			

We will use the VLOOKUP function to do this.

Into C2 type, ' =
VLOOKUP(A2,Sheet1!$A$2:$B$6, 2,
FALSE) ' then press enter. Sheet1! tells
VLOOKUP to go to Sheet1 to find the range
$A$2:$B$6. The result is the value $35
then use the handle to complete the table.

	A	B	C
	G37		f_x
1	Name	Hours	Hourly Rate
2	Diane	42	$35
3	Ed	36	$35
4	Alan	43	$40
5	Bob	29	$45
6	Cathy	56	$50

Before we began it was necessary to make
sure that a check was made that the data in
the lookup column was in ascending order.
As it was, there was no problem.

We can easily calculate the week's pay of
the employees by typing, '= B2*C2', into D2
then pressing enter followed by the handle.

G8		fx	
A	B	C	D
1 Name	Hours	Hourly Rate	Week's Pay
2 Diane	42	$35	$1,470
3 Ed	36	$35	$1,260
4 Alan	43	$40	$1,720
5 Bob	29	$45	$1,305
6 Cathy	56	$50	$2,800

Before finishing this function, let us examine the last parameter, which takes the value FALSE for an exact match and TRUE for an approximate match.

Leave the function as it is but change Diane to Diana. The table below shows what happens.

C9		fx	
A	B	C	D
1 Name	Hours	Hourly Rate	Week's Pay
2 Diana	42	#N/A	#N/A
3 Ed	36	$35	$1,260
4 Alan	43	$40	$1,720
5 Bob	29	$45	$1,305
6 Cathy	56	$50	$2,800

However, now change FALSE to TRUE in the function declaration, ' = VLOOKUP(A2, Sheet1!$A$2:$B$6, 2, TRUE) ' and then press enter followed by the handle.

	A	B	C	D
1	Name	Hours	Hourly Rate	Week's Pay
2	Diana	42	$50	$2,100
3	Ed	36	$35	$1,260
4	Alan	43	$40	$1,720
5	Bob	29	$45	$1,305
6	Cathy	56	$50	$2,800

VLOOKUP has assigned Cathy's hourly rate, probably because the names Cathy and Diana have the same number of characters or letters. Cathy and Diana are regarded as approximately equal.

Finally, we have a look at the syntax of the VLOOKUP function. As usual, a declaration is very useful: ' = VLOOKUP(H2,$A$2:$E$21, 5, FALSE) '.

The parameters in order are, a cell [where the value to be looked up is], the range where the function looks, the number of the column where the value is found and a Boolean value of TRUE or FALSE.

Chapter 19: HLOOKUP

VLOOKUP used data arranged vertically; HLOOKUP uses data arranged horizontally.

How can data be arranged horizontally? Here is an example:

A8			fx	
A	B	C	D	E
1	Make	Ford	Toyota	Honda	Chrysler
2	In Stock	20	23	11	5
3	Last Sale	May 3,2019	May 2, 2019	May 3, 2019	May 1, 2019

Here is how you could access this simple table using HLOOKUP. The problem is to find the number of Honda cars in stock, using the HLOOKUP function, and write it in the cell B6 to the right of the word Honda in cell A6.

6	Honda

In cell B6, type the function declaration,'
=HLOOKUP("Honda",A1:E3, 2, FALSE)'
then press enter.

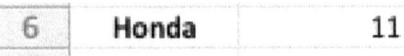

| 6 | Honda | 11 |

We get the same result if we have '
=HLOOKUP(A6, A1: E3, 2, FALSE)'. The
reason is that the word 'Honda' is
contained in A6.

However, see what happens if have '
=HLOOKUP(A2, A1: E3, 2, FALSE)'.

| 6 | Honda | #N/A |

The reason is that the word 'Honda' is not
contained in A2.

Now, see what happens if have '
=HLOOKUP("FORD", A1: E3, 2, FALSE)'.

| 6 | Honda | 20 |

The reason is that the word 'Ford' is contained in the top row of the table with 20 below it in row 2.

As with VLOOKUP, we need to examine the function declaration and explain what is happening. Here is the original function declaration, ' =HLOOKUP("Honda", A1:E3, 2, FALSE)'.

HLOOKUP searches for the word Honda in the **top row** of the range A1: E3. Once this is found, the 2 tells HLOOKUP to take the contents of row 2, which is beneath Honda and put it in B6. The FALSE tells HLOOKUP that there must be an exact match. An approximate match, such as HINDA or HONDO would not be acceptable, whereas it might for TRUE.

Finally, you may recall that VLOOKUP required the left column to be in ascending order. For HLOOKUP, this is not necessary for a last parameter value of FALSE. However, it is necessary that the top row of the range be in ascending order if the last parameter is TRUE.

We have often finished chapters on functions with a paragraph about structure or syntax. We will not dwell on structure or syntax. The syntax is very similar to that for VLOOKUP. If you are interested then see what was written in that chapter about this topic.

Chapter 20: A few notes about Pasting, Ordering and Filtering

Usually, basic information such as this is covered early. However, the notes about ordering are only vital when you use LOOKUP functions.

Pasting

Here is the result of multiplying two columns of numbers using the declaration, '= **A2*B2**' then the handle.

	C2					fx	=A2*B2
	A		B		C		D
1							
2		3		4		12	
3		7		9		63	
4		5		6		30	
5		4		5		20	
6		6		9		54	

Often having computed a column like this it might be useful to eliminate one or both of the source columns but if you delete them this is what happens.

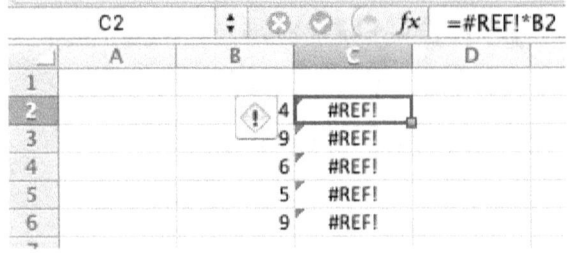

	A	B	C	D
1				
2		4	#REF!	
3		9	#REF!	
4		6	#REF!	
5		5	#REF!	
6		9	#REF!	

To avoid this use Paste Values

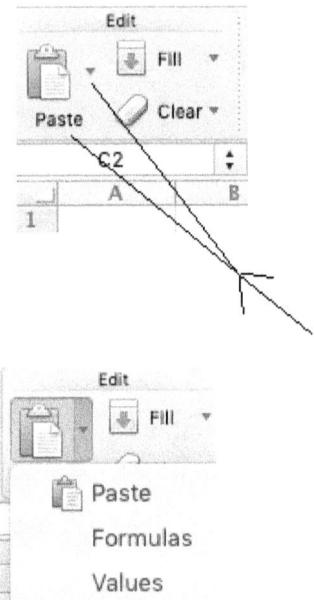

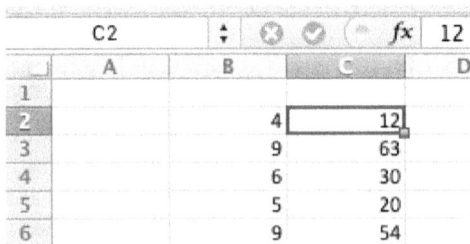

after copying the original.

	A	B	C	D
1				
2		4	12	
3		9	63	
4		6	30	
5		5	20	
6		9	54	

Now after deleting one the table of products is unaffected, as can be seen by looking at the diagram above.

Ordering
As mentioned and emphasized in the chapter on VLOOKUP, it is absolutely important that the first column is in ascending order. This is all well and good but how is this done?

Suppose the range we are going to look up from is the table below.

	A	B	C	D	E
	ID	Family	First	Gender	Wage per hour
1					
2	B45	Benton	Jimmy	M	$25
3	G06	Brooke	Helen	F	$25
4	A56	Cheong	Greta	F	$23
5	H67	Griffith	Murgatroyd	M	$30
6	A01	Hicke	Sue	F	$25
7	A12	Howan	June	F	$67
8	H02	Jorgensen	Fred	M	$32
9	H13	McAndrew	Diana	F	$45
10	G22	Mildeng	Zane	M	$56
11	F11	Pang	Peter	M	$25
12					

and suppose the lookup is based on ID. You can see that the table has been ordered by Family.

To order this range by ID is very easy.

Here are the steps, after you have selected the range A2: E11.

A2		fx	B45	
A	**B**	**C**	**D**	**E**
ID	Family	First	Gender	Wage per hour
B45	Benton	Jimmy	M	$25
G06	Brooke	Helen	F	$25
A56	Cheong	Greta	F	$23
H67	Griffith	Murgatroyd	M	$30
A01	Hicke	Sue	F	$25
A12	Howan	June	F	$67
H02	Jorgensen	Fred	M	$32
H13	McAndrew	Diana	F	$45
G22	Mildeng	Zane	M	$56
F11	Pang	Peter	M	$25

(1) Click on the DATA tab on your Excel Menu.

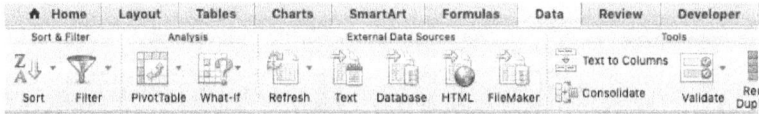

(2) Open the Sort & Filter window.

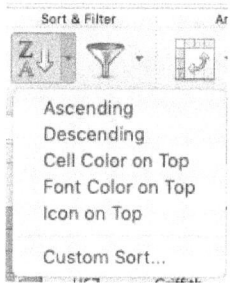

(3) Pick Custom Sort [I always use this except for trivial sorts]

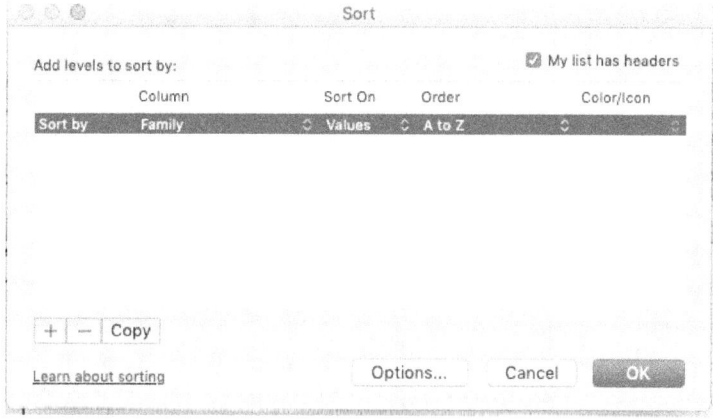

(4) Click on Family

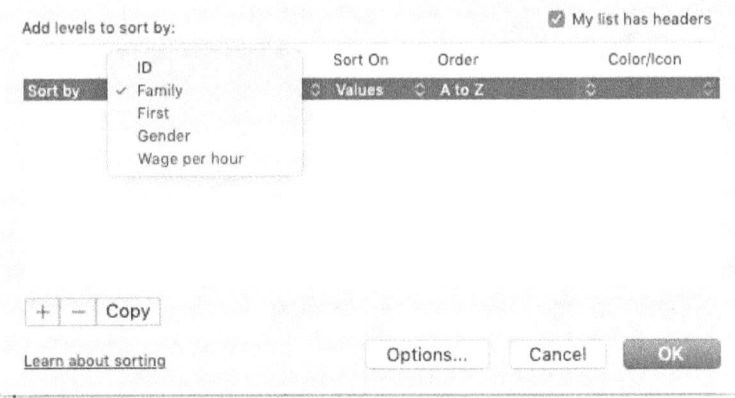

Add levels to sort by:				My list has headers
	ID	Sort On	Order	Color/Icon
Sort by	✓ Family	Values	A to Z	
	First			
	Gender			
	Wage per hour			

+	−	Copy

Learn about sorting   Options...   Cancel   **OK**

(5) Select ID and press OK.

| A2 | | ⊗ ✓ ⌢ | *fx* | A01 |

	A	B	C	D	E
1	ID	Family	First	Gender	Wage per hour
2	A01	Hicke	Sue	F	$25
3	A12	Howan	June	F	$67
4	A56	Cheong	Greta	F	$23
5	B45	Benton	Jimmy	M	$25
6	F11	Pang	Peter	M	$25
7	G06	Brooke	Helen	F	$25
8	G22	Mildeng	Zane	M	$56
9	H02	Jorgensen	Fred	M	$32
10	H13	McAndrew	Diana	F	$45
11	H67	Griffith	Murgatroyd	M	$30

The VLOOKUP function can now be used.

Further Ordering

Consider the following data set of student results from a high school.

	A	B	C	D	E	F	G
1	ID	Family	First	Gender	English	Math	Science
2	A67	Argentine	Brad	M	47	64	88
3	M78	Chadhaman	Dipsya	F	61	42	57
4	D57	Davison	Aimee	F	61	57	50
5	A98	Gleeson	Wallace	M	50	28	42
6	F56	Greenwell	Fiona	F	53	92	57
7	B01	Hrstich	Jack	M	57	23	57
8	Y67	Huang	Stephani	F	28	57	51
9	A21	Jitney	Steve	M	59	66	46
10	C34	Kennedy	Mary	F	81	87	47
11	F82	Leotard	Tereise	F	57	81	53
12	Y76	Li	Jill	F	88	77	58
13	B98	Motlagh	Sam	M	78	14	71
14	F83	Pecovera	Ellen	F	51	64	78
15	H25	Randal	Sophie	F	49	46	61
16	H03	Raufger	Shazinat	F	58	56	81
17	A09	Shao	Bill	M	71	45	78
18	Z07	Sidwell	Amelia	F	45	71	42
19	B67	Smith	Jos	M	53	33	77
20	H46	Sproully	Larissa	F	85	88	53
21	H23	Sung	Kate	F	42	78	81
22	C02	Zumpar	Don	M	31	21	57

We want the results of the female students only with a total over the three subjects and the results put in descending order, with the highest total at the top, down to the lowest.

First, let's get the totals.

Into H2, type the formula declaration, **' =
E2 + F2 + G2'** then press equal. Once this
is done use the handle to get the data set
below.

	A	B	C	D	E	F	G	H
	ID	Family	First	Gender	English	Math	Science	Total
1								
2	A67	Argentine	Brad	M	47	64	88	199
3	M78	Chadhaman	Dipsya	F	61	42	57	160
4	D57	Davison	Aimee	F	61	57	50	168
5	A98	Gleeson	Wallace	M	50	28	42	120
6	F56	Greenwell	Fiona	F	53	92	57	202
7	B01	Hrstich	Jack	M	57	23	57	137
8	Y67	Huang	Stephani	F	28	57	51	136
9	A21	Jitney	Steve	M	59	66	46	171
10	C34	Kennedy	Mary	F	81	87	47	215
11	F82	Leotard	Tereise	F	57	81	53	191
12	Y76	Li	Jill	F	88	77	58	223
13	B98	Motlagh	Sam	M	78	14	71	163
14	F83	Pecovera	Ellen	F	51	64	78	193
15	H25	Randal	Sophie	F	49	46	61	156
16	H03	Raufger	Shazinat	F	58	56	81	195
17	A09	Shao	Bill	M	71	45	78	194
18	Z07	Sidwell	Amelia	F	45	71	42	158
19	B67	Smith	Jos	M	53	33	77	163
20	H46	Sproully	Larissa	F	85	88	53	226
21	H23	Sung	Kate	F	42	78	81	201
22	C02	Zumpar	Don	M	31	21	57	109

Now select A1: H22 and go to Data.

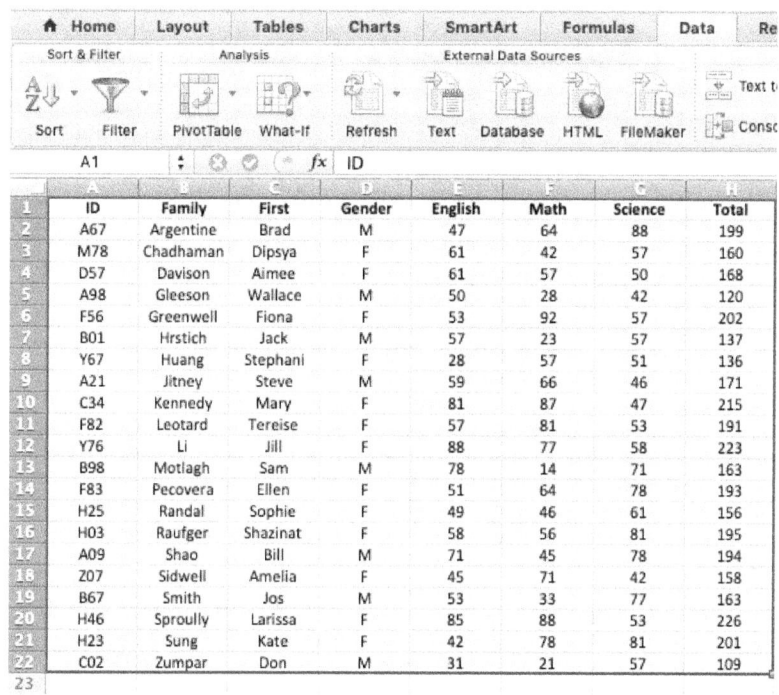

	ID	Family	First	Gender	English	Math	Science	Total
1	ID	Family	First	Gender	English	Math	Science	Total
2	A67	Argentine	Brad	M	47	64	88	199
3	M78	Chadhaman	Dipsya	F	61	42	57	160
4	D57	Davison	Aimee	F	61	57	50	168
5	A98	Gleeson	Wallace	M	50	28	42	120
6	F56	Greenwell	Fiona	F	53	92	57	202
7	B01	Hrstich	Jack	M	57	23	57	137
8	Y67	Huang	Stephani	F	28	57	51	136
9	A21	Jitney	Steve	M	59	66	46	171
10	C34	Kennedy	Mary	F	81	87	47	215
11	F82	Leotard	Tereise	F	57	81	53	191
12	Y76	Li	Jill	F	88	77	58	223
13	B98	Motlagh	Sam	M	78	14	71	163
14	F83	Pecovera	Ellen	F	51	64	78	193
15	H25	Randal	Sophie	F	49	46	61	156
16	H03	Raufger	Shazinat	F	58	56	81	195
17	A09	Shao	Bill	M	71	45	78	194
18	Z07	Sidwell	Amelia	F	45	71	42	158
19	B67	Smith	Jos	M	53	33	77	163
20	H46	Sproully	Larissa	F	85	88	53	226
21	H23	Sung	Kate	F	42	78	81	201
22	C02	Zumpar	Don	M	31	21	57	109

Go to Sort & Filter and do a custom sort based on Gender. Here is what you should have.

	A	B	C	D	E	F	G	H
1	ID	Family	First	Gender	English	Math	Science	Total
2	M78	Chadhaman	Dipsya	F	61	42	57	160
3	D57	Davison	Aimee	F	61	57	50	168
4	F56	Greenwell	Fiona	F	53	92	57	202
5	Y67	Huang	Stephani	F	28	57	51	136
6	C34	Kennedy	Mary	F	81	87	47	215
7	F82	Leotard	Tereise	F	57	81	53	191
8	Y76	Li	Jill	F	88	77	58	223
9	F83	Pecovera	Ellen	F	51	64	78	193
10	H25	Randal	Sophie	F	49	46	61	156
11	H03	Raufger	Shazinat	F	58	56	81	195
12	Z07	Sidwell	Amelia	F	45	71	42	158
13	H46	Sproully	Larissa	F	85	88	53	226
14	H23	Sung	Kate	F	42	78	81	201
15	A67	Argentine	Brad	M	47	64	88	199
16	A98	Gleeson	Wallace	M	50	28	42	120
17	B01	Hrstich	Jack	M	57	23	57	137
18	A21	Jitney	Steve	M	59	66	46	171
19	B98	Motlagh	Sam	M	78	14	71	163
20	A09	Shao	Bill	M	71	45	78	194
21	B67	Smith	Jos	M	53	33	77	163
22	C02	Zumpar	Don	M	31	21	57	109

Now select the range A1: H14, which is the range with females and copy this range [ctrl-c in a PC or command-c in a Mac]. Paste in some place, using a special paste with Values. The result is shown below.

ID	Family	First	Gender	English	Math	Science	Total
M78	Chadhaman	Dipsya	F	61	42	57	160
D57	Davison	Aimee	F	61	57	50	168
F56	Greenwell	Fiona	F	53	92	57	202
Y67	Huang	Stephani	F	28	57	51	136
C34	Kennedy	Mary	F	81	87	47	215
F82	Leotard	Tereise	F	57	81	53	191
Y76	Li	Jill	F	88	77	58	223
F83	Pecovera	Ellen	F	51	64	78	193
H25	Randal	Sophie	F	49	46	61	156
H03	Raufger	Shazinat	F	58	56	81	195
Z07	Sidwell	Amelia	F	45	71	42	158
H46	Sproully	Larissa	F	85	88	53	226
H23	Sung	Kate	F	42	78	81	201

Now we need to sort them via total in descending order. Once again select the range and go to Sort and Filter.

Select Custom Sort and make sure you pick Total and Largest to Smallest.

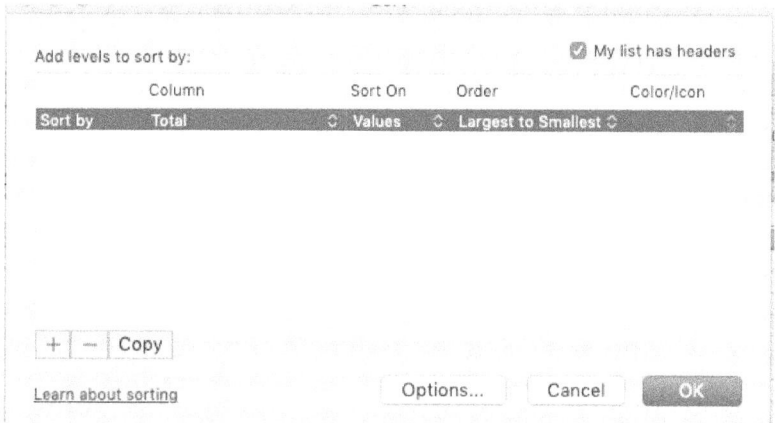

We now have the desired table.

ID	Family	First	Gender	English	Math	Science	Total
H46	Sproully	Larissa	F	85	88	53	226
Y76	Ii	Jill	F	88	77	58	223
C34	Kennedy	Mary	F	81	87	47	215
F56	Greenwell	Fiona	F	53	92	57	202
H23	Sung	Kate	F	42	78	81	201
H03	Raufger	Shazinat	F	58	56	81	195
F83	Pecovera	Ellen	F	51	64	78	193
F82	Leotard	Tereise	F	57	81	53	191
D57	Davison	Aimee	F	61	57	50	168
M78	Chadhaman	Dipsya	F	61	42	57	160
Z07	Sidwell	Amelia	F	45	71	42	158
H25	Randal	Sophie	F	49	46	61	156
Y67	Huang	Stephani	F	28	57	51	136

Some of you may have wondered about Filter. What is this? Read on to find out.

Filter
I will not spend too much time on this. Basically, it is a way of examining a set of data using constraints but leaving the data set intact.

Here is an example. The data set below is the same one we extracted the females from before using order. This time we are going to have a look at the male results in ascending order of Total [smallest to largest], save the filtered data set elsewhere then restore the data set.

	A	B	C	D	E	F	G	H
1	ID	Family	First	Gender	English	Math	Science	Total
2	H46	Sproully	Larissa	F	85	88	53	226
3	Y76	Li	Jill	F	88	77	58	223
4	C34	Kennedy	Mary	F	81	87	47	215
5	F56	Greenwell	Fiona	F	53	92	57	202
6	H23	Sung	Kate	F	42	78	81	201
7	H03	Raufger	Shazinat	F	58	56	81	195
8	A09	Shao	Bill	M	71	45	78	194
9	F83	Pecovera	Ellen	F	51	64	78	193
10	F82	Leotard	Tereise	F	57	81	53	191
11	A21	Jitney	Steve	M	59	66	46	171
12	D57	Davison	Aimee	F	61	57	50	168
13	B98	Motlagh	Sam	M	78	14	71	163
14	B67	Smith	Jos	M	53	33	77	163
15	M78	Chadhaman	Dipsya	F	61	42	57	160
16	Z07	Sidwell	Amelia	F	45	71	42	158
17	A67	Argentine	Brad	M	47	64	88	199
18	H25	Randal	Sophie	F	49	46	61	156
19	B01	Hrstich	Jack	M	57	23	57	137
20	Y67	Huang	Stephani	F	28	57	51	136
21	A98	Gleeson	Wallace	M	50	28	42	120
22	C02	Zumpar	Don	M	31	21	57	109
23								
24								

Go to the Sort & Filter tab, as you did
before, after clicking on the Data tab of the
Excel menu.

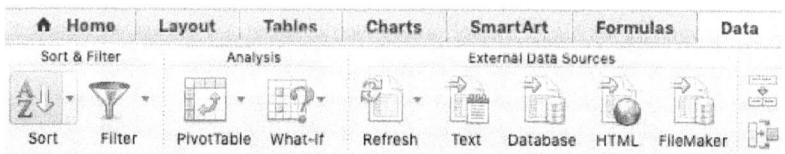

This time click Filter, but NOT the down
arrow, after selecting the whole data set.

	A	B	C	D	E	F	G	H
1	ID ▼	Family ▼	First ▼	Gender ▼	English ▼	Math ▼	Science ▼	Total ▼
2	H46	Sproully	Larissa	F	85	88	53	226
3	Y76	Li	Jill	F	88	77	58	223
4	C34	Kennedy	Mary	F	81	87	47	215
5	F56	Greenwell	Fiona	F	53	92	57	202
6	H23	Sung	Kate	F	42	78	81	201
7	H03	Raufger	Shazinat	F	58	56	81	195
8	A09	Shao	Bill	M	71	45	78	194
9	F83	Pecovera	Ellen	F	51	64	78	193
10	F82	Leotard	Tereise	F	57	81	53	191
11	A21	Jitney	Steve	M	59	66	46	171
12	D57	Davison	Aimee	F	61	57	50	168
13	B98	Motlagh	Sam	M	78	14	71	163
14	B67	Smith	Jos	M	53	33	77	163
15	M78	Chadhaman	Dipsya	F	61	42	57	160
16	Z07	Sidwell	Amelia	F	45	71	42	158
17	A67	Argentine	Brad	M	47	64	88	199
18	H25	Randal	Sophie	F	49	46	61	156
19	B01	Hrstich	Jack	M	57	23	57	137
20	Y67	Huang	Stephani	F	28	57	51	136
21	A98	Gleeson	Wallace	M	50	28	42	120
22	C02	Zumpar	Don	M	31	21	57	109
23								
24								

You will notice little arrows. Click on the arrow in the Gender column.

Immediately you do, the following window appears.

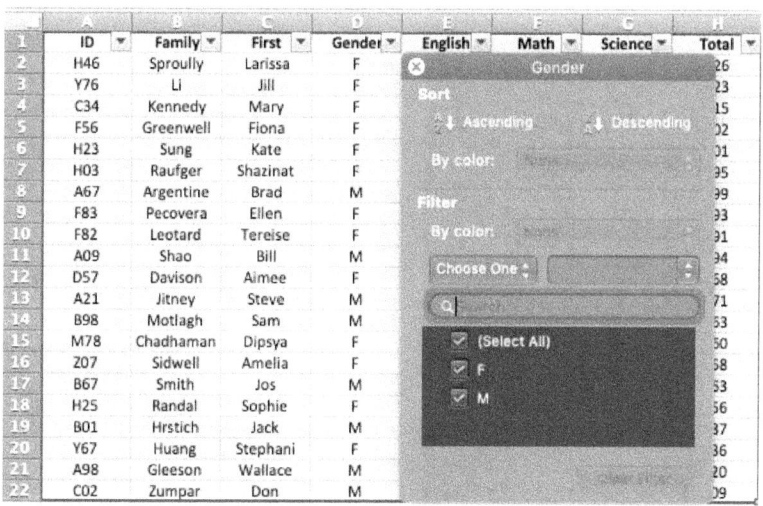

Fill it in as shown.

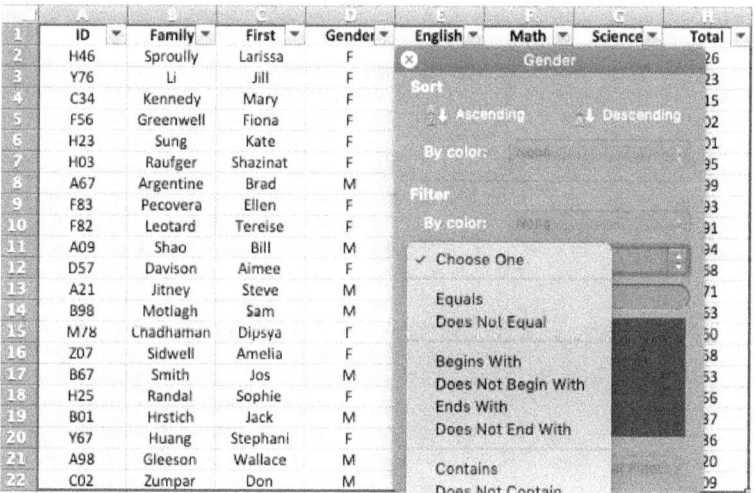

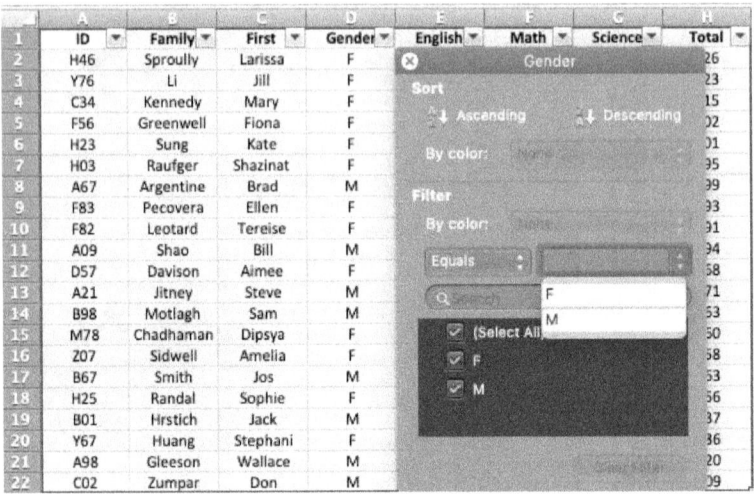

Leading to

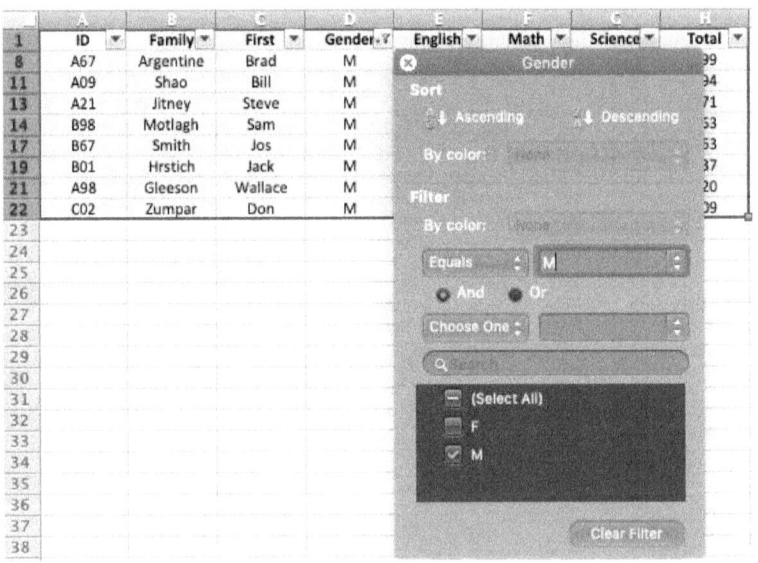

Finally, close the Gender window and you only have the Male results.

	A	B	C	D	E	F	G	H
1	ID ▾	Family ▾	First ▾	Gender ▾	English ▾	Math ▾	Science ▾	Total ▾
8	A67	Argentine	Brad	M	47	64	88	199
11	A09	Shao	Bill	M	71	45	78	194
13	A21	Jitney	Steve	M	59	66	46	171
14	B98	Motlagh	Sam	M	78	14	71	163
17	B67	Smith	Jos	M	53	33	77	163
19	B01	Hrstich	Jack	M	57	23	57	137
21	A98	Gleeson	Wallace	M	50	28	42	120
22	C02	Zumpar	Don	M	31	21	57	109

These results are in descending order. However, as it is necessary to put them in ascending order, you use the little arrow beside Total. A little window appears in which ascending is an option.

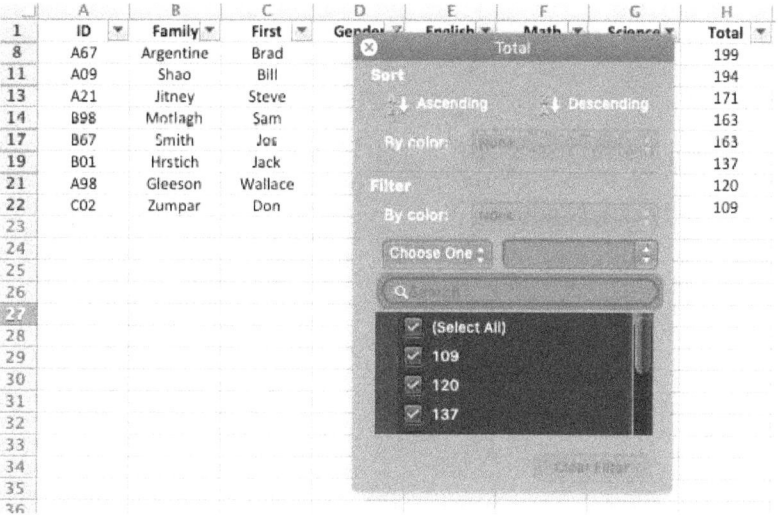

Clicking on this will put the results in ascending order.

	A	B	C	D	E	F	G	H
	K27			fx				
1	ID ▼	Family ▼	First ▼	Gender-▼	English ▼	Math ▼	Science ▼	Total ▼↑
8	C02	Zumpar	Don	M	31	21	57	109
11	A98	Gleeson	Wallace	M	50	28	42	120
13	B01	Hrstich	Jack	M	57	23	57	137
14	B98	Motlagh	Sam	M	78	14	71	163
17	B67	Smith	Jos	M	53	33	77	163
19	A21	Jitney	Steve	M	59	66	46	171
21	A09	Shao	Bill	M	71	45	78	194
22	A67	Argentine	Brad	M	47	64	88	199

If you want to save this as a data set then just select the set and paste it somewhere. I have pasted it just below the filter. Notice that the data is copied as Values.

	A	B	C	D	E	F	G	H
	H25			fx	109			
1	ID ▼	Family ▼	First ▼	Gender-▼	English ▼	Math ▼	Science ▼	Total ▼↑
8	C02	Zumpar	Don	M	31	21	57	109
11	A98	Gleeson	Wallace	M	50	28	42	120
13	B01	Hrstich	Jack	M	57	23	57	137
14	B98	Motlagh	Sam	M	78	14	71	163
17	B67	Smith	Jos	M	53	33	77	163
19	A21	Jitney	Steve	M	59	66	46	171
21	A09	Shao	Bill	M	71	45	78	194
22	A67	Argentine	Brad	M	47	64	88	199
23								
24	ID	Family	First	Gender	English	Math	Science	Total
25	C02	Zumpar	Don	M	31	21	57	109
26	A98	Gleeson	Wallace	M	50	28	42	120
27	B01	Hrstich	Jack	M	57	23	57	137
28	B98	Motlagh	Sam	M	78	14	71	163
29	B67	Smith	Jos	M	53	33	77	163
30	A21	Jitney	Steve	M	59	66	46	171
31	A09	Shao	Bill	M	71	45	78	194
32	A67	Argentine	Brad	M	47	64	88	199

Finally, restoring the original set of data is easy. Just click on the Filter icon. When you do this the Filter arrows are removed and you have the original data set restored.

H25			fx	109			
A	B	C	D	E	F	G	
ID	Family	First	Gender	English	Math	Science	Total
H46	Sproully	Larissa	F	85	88	53	226
Y76	Li	Jill	F	88	77	58	223
C34	Kennedy	Mary	F	81	87	47	215
F56	Greenwell	Fiona	F	53	92	57	202
H23	Sung	Kate	F	42	78	81	201
H03	Raufger	Shazinat	F	58	56	81	195
C02	Zumpar	Don	M	31	21	57	109
F83	Pecovera	Ellen	F	51	64	78	193
F82	Leotard	Tereise	F	57	81	53	191
A98	Gleeson	Wallace	M	50	28	42	120
D57	Davison	Aimee	F	61	57	50	168
B01	Hrstich	Jack	M	57	23	57	137
B98	Motlagh	Sam	M	78	14	71	163
M78	Chadhaman	Dipsya	F	61	42	57	160
Z07	Sidwell	Amelia	F	45	71	42	158
B67	Smith	Jos	M	53	33	77	163
H25	Randal	Sophie	F	49	46	61	156
A21	Jitney	Steve	M	59	66	46	171
Y67	Huang	Stephani	F	28	57	51	136
A09	Shao	Bill	M	71	45	78	194
A67	Argentine	Brad	M	47	64	88	199

Filtering by Color
While exploring Filter, you may have noticed the option Filter by Cell Color.

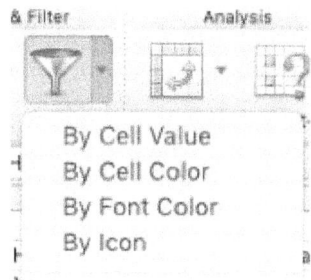

Hitherto, this has been irrelevant, as we have only used Black and White. However, Excel offers all sorts of color schemes. Two simple ones for Fill Color and Font Color are shown.

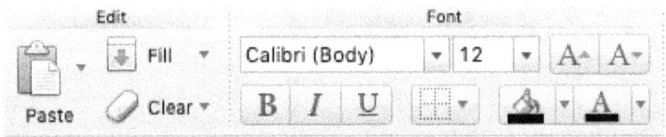

To show the use of Filter by Cell Color, we have colored the data set we have been using.

	H35			fx				
	A	B	C	D	E	F	G	H
1	ID	Family	First	Gender	English	Math	Science	Total
2	H46	Sproully	Larissa	F	85	88	53	226
3	Y76	Li	Jill	F	88	77	58	223
4	C34	Kennedy	Mary	F	81	87	47	215
5	F56	Greenwell	Fiona	F	53	92	57	202
6	H23	Sung	Kate	F	42	78	81	201
7	H03	Raufger	Shazinat	F	58	56	81	195
8	C02	Zumpar	Don	M	31	21	57	109
9	F83	Pecovera	Ellen	F	51	64	78	193
10	F82	Leotard	Tereise	F	57	81	53	191
11	A98	Gleeson	Wallace	M	50	28	42	120
12	D57	Davison	Aimee	F	61	57	50	168
13	B01	Hrstich	Jack	M	57	23	57	137
14	B98	Motlagh	Sam	M	78	14	71	163
15	M78	Chadhaman	Dipsya	F	61	42	57	160
16	Z07	Sidwell	Amelia	F	45	71	42	158
17	B67	Smith	Jos	M	53	33	77	163
18	H25	Randal	Sophie	F	49	46	61	156
19	A21	Jitney	Steve	M	59	66	46	171
20	Y67	Huang	Stephani	F	28	57	51	136
21	A09	Shao	Bill	M	71	45	78	194
22	A67	Argentine	Brad	M	47	64	88	199

Now click on Filter after selecting the data set. This results in the following.

	L29			fx				
	A	B	C	D	E	F	G	H
1	ID ▼	Family ▼	First ▼	Gender ▼	English ▼	Math ▼	Science ▼	Total ▼
2	H46	Sproully	Larissa	F	85	88	53	226
3	Y76	Li	Jill	F	88	77	58	223
4	C34	Kennedy	Mary	F	81	87	47	215
5	F56	Greenwell	Fiona	F	53	92	57	202
6	H23	Sung	Kate	F	42	78	81	201
7	H03	Raufger	Shazinat	F	58	56	81	195
8	C02	Zumpar	Don	M	31	21	57	109
9	F83	Pecovera	Ellen	F	51	64	78	193
10	F82	Leotard	Tereise	F	57	81	53	191
11	A98	Gleeson	Wallace	M	50	28	42	120
12	D57	Davison	Aimee	F	61	57	50	168
13	B01	Hrstich	Jack	M	57	23	57	137
14	B98	Motlagh	Sam	M	78	14	71	163
15	M78	Chadhaman	Dipsya	F	61	42	57	160
16	Z07	Sidwell	Amelia	F	45	71	42	158
17	B67	Smith	Jos	M	53	33	77	163
18	H25	Randal	Sophie	F	49	46	61	156
19	A21	Jitney	Steve	M	59	66	46	171
20	Y67	Huang	Stephani	F	28	57	51	136
21	A09	Shao	Bill	M	71	45	78	194
22	A67	Argentine	Brad	M	47	64	88	199
23								

Now use any little arrow, choose color and
finally yellow or red.

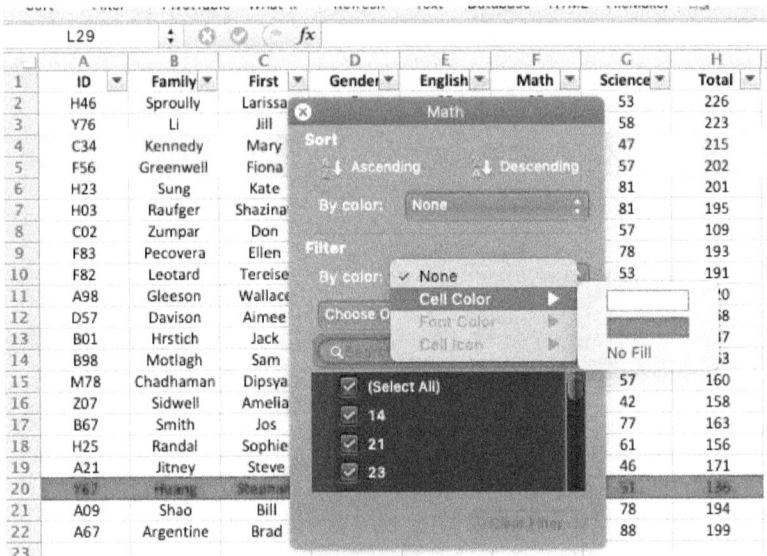

If we pick red then we get the filter below.

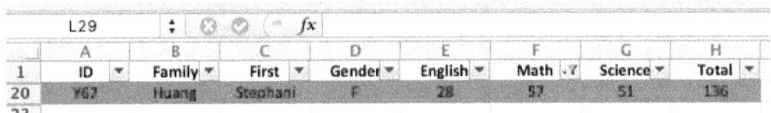

We restore the original by clicking on the Filter icon.

	A	B	C	D	E	F	G	H
1	ID	Family	First	Gender	English	Math	Science	Total
2	H46	Sproully	Larissa	F	85	88	53	226
3	Y76	Li	Jill	F	88	77	58	223
4	C34	Kennedy	Mary	F	81	87	47	215
5	F56	Greenwell	Fiona	F	53	92	57	202
6	H23	Sung	Kate	F	42	78	81	201
7	H03	Raufger	Shazinat	F	58	56	81	195
8	C02	Zumpar	Don	M	31	21	57	109
9	F83	Pecovera	Ellen	F	51	64	78	193
10	F82	Leotard	Tereise	F	57	81	53	191
11	A98	Gleeson	Wallace	M	50	28	42	120
12	D57	Davison	Aimee	F	61	57	50	168
13	B01	Hrstich	Jack	M	57	23	57	137
14	B98	Motlagh	Sam	M	78	14	71	163
15	M78	Chadhaman	Dipsya	F	61	42	57	160
16	Z07	Sidwell	Amelia	F	45	71	42	158
17	B67	Smith	Jos	M	53	33	77	163
18	H25	Randal	Sophie	F	49	46	61	156
19	A21	Jitney	Steve	M	59	66	46	171
20	Y67	Huang	Stephani	F	28	57	51	136
21	A09	Shao	Bill	M	71	45	78	194
22	A67	Argentine	Brad	M	47	64	88	199

Conclusion

Excel has a computer language called VBA [Visual Basic for Applications] associated with it. Really good exponents of Excel use VBA all the time

If you wish to master it you must know the basics of formulas and functions inside and out.

This book has thoroughly covered many of those basics.

Before you proceed further with formulas and functions you must master the material in the book to the extent that you can do the problems in every chapter.

Once you have mastered the material in this book you are ready for all the other things that can be done with EXCEL and the VBA programming language!

Good Luck.

Welcome to the last page reader, I'm happy to see you here I hope you had a great time reading my book and if you want to support my work, leaving an honest review will be highly appreciated. Thank you so much!

Respectfully,
William B. Skates